Smithsonian

SPACECRAFT

Coloring Book

ILLUSTRATED BY
John Pirtel

IDW

Facebook: **facebook.com/idwpublishing**
Twitter: **@idwpublishing**
YouTube: **youtube.com/idwpublishing**
Instagram: **@idwpublishing**

ISBN : 978-1-68405-828-0   21 22 23 24   4 3 2 1

TEXT BY
**SMITHSONIAN'S NATIONAL
AIR & SPACE MUSEUM**

ILLUSTRATIONS BY
JOHN PIRTEL

CREATIVE ASSISTANCE BY
ERIK AUSTIN

EDITED BY
JUSTIN EISINGER

EDITORIAL ASSISTANCE BY
EDWARD GAUVIN

BOOK DESIGN BY
NATHAN WIDICK

Nachie Marsham, Publisher
Rebekah Cahalin, EVP of Operations
Blake Kobashigawa, VP of Sales
John Barber, Editor-in-Chief
Mark Doyle, Editorial Director, Originals
Justin Eisinger, Editorial Director, Graphic Novels and Collections
Scott Dunbier, Director, Special Projects
Anna Morrow, Sr. Marketing Director
Tara McCrillis, Director of Design & Production
Shauna Monteforte, Sr. Director of Manufacturing Operations

Ted Adams and Robbie Robbins, IDW Founders

Special thanks to the team at The Smithsonian for all of their assistance and support.

Smithsonian Enterprises:
Kealy Gordon, Product Development Manager
Jill Corcoran, Director, Licensed Publishing
Janet Archer, DMM, Ecom and D-to-C
Carol LeBlanc, President

Smithsonian's National Air & Space Museum:
Martin Collins, Curator
James David, Curator
David DeVorkin, Senior Curator
Jennifer Levasseur, Curator
Cathleen Lewis, Curator

Teasel Muir-Harmony, Curator
Michael J. Neufeld, Senior Curator
Matthew Shindell, Curator
Margaret A. Weitekamp, Curator

The Smithsonian's National Air and Space Museum maintains the world's largest and most significant collection of aviation and space artifacts, encompassing all aspects of human flight, as well as related works of art and archival materials. It operates two landmark facilities that, together, welcome more than eight million visitors a year, making it the most visited museum in the country. It also is home to the Center for Earth and Planetary Studies.

The publisher would like to thank The National Aeronautics and Space Administration, The National Oceanic and Atmospheric Administration, *The New York Times*, The Stafford Air & Space Museum, Spacefacts, Astronautix, Skyrocket, Cradle Of Aviation, The Coca-Cola Space Science Center, and Boeing for providing additional information.

This Book Belongs To:

_____

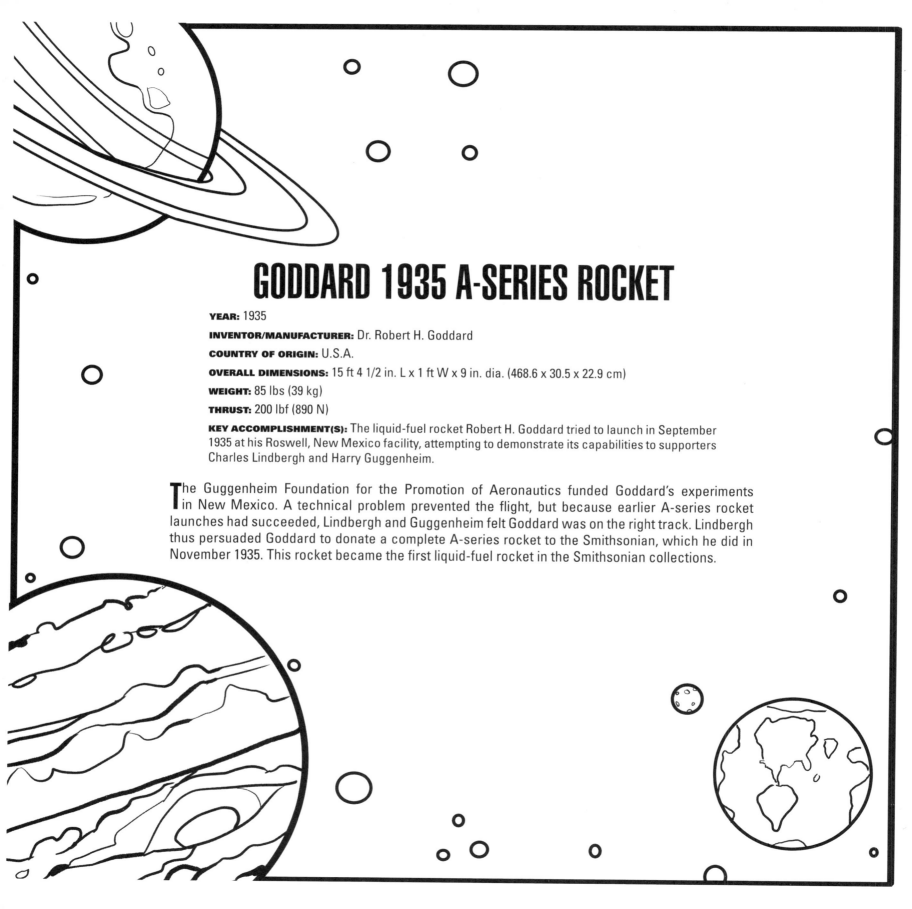

GODDARD 1935 A-SERIES ROCKET

YEAR: 1935

INVENTOR/MANUFACTURER: Dr. Robert H. Goddard

COUNTRY OF ORIGIN: U.S.A.

OVERALL DIMENSIONS: 15 ft 4 1/2 in. L x 1 ft W x 9 in. dia. (468.6 x 30.5 x 22.9 cm)

WEIGHT: 85 lbs (39 kg)

THRUST: 200 lbf (890 N)

KEY ACCOMPLISHMENT(S): The liquid-fuel rocket Robert H. Goddard tried to launch in September 1935 at his Roswell, New Mexico facility, attempting to demonstrate its capabilities to supporters Charles Lindbergh and Harry Guggenheim.

The Guggenheim Foundation for the Promotion of Aeronautics funded Goddard's experiments in New Mexico. A technical problem prevented the flight, but because earlier A-series rocket launches had succeeded, Lindbergh and Guggenheim felt Goddard was on the right track. Lindbergh thus persuaded Goddard to donate a complete A-series rocket to the Smithsonian, which he did in November 1935. This rocket became the first liquid-fuel rocket in the Smithsonian collections.

WAC CORPORAL SOUNDING ROCKET

YEAR: 1945

INVENTOR/MANUFACTURER: Jet Propulsion Laboratory (JPL), California Institute of Technology

COUNTRY OF ORIGIN: U.S.A.

OVERALL DIMENSIONS: 3 ft W x 16 ft L x 1 ft dia. (91.4 x 487.7 x 30.5 cm)

WEIGHT: 292 lbs (132.5 kg)

THRUST: Tiny Tim booster 50,000 lbf (222,411 N); WAC Corporal sustainer 1,500 lbf (6,672.3 N)

KEY ACCOMPLISHMENT(S): The U.S.'s first successful sounding rocket, or research rocket, instrument-carrying rockets designed to take measurements and perform scientific experiments during sub-orbital flight.

The initials WAC have been variously said to stand for "Without Attitude Control" or "Women's Army Corps." Developed from 1944 onwards at the Jet Propulsion Laboratory, the WAC Corporal could lift 25 pounds of instruments to 20 miles with its nitric acid and aniline motor. The first rocket was launched in 1945. However, captured German V-2 rockets soon became available that could carry heavier payloads to higher altitudes. The WAC was thus little used. One was placed on top of a V-2, however, as part of the U.S.'s first experimental two-stage liquid propellant rocket series called Project Bumper.

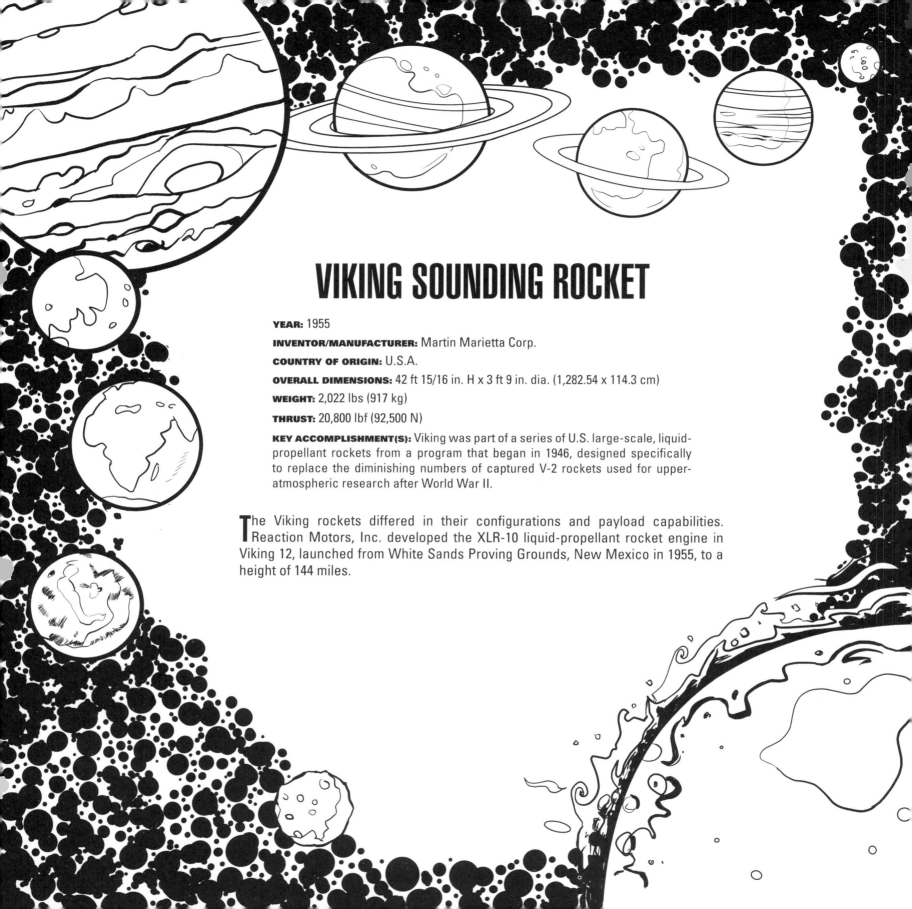

VIKING SOUNDING ROCKET

YEAR: 1955

INVENTOR/MANUFACTURER: Martin Marietta Corp.

COUNTRY OF ORIGIN: U.S.A.

OVERALL DIMENSIONS: 42 ft 15/16 in. H x 3 ft 9 in. dia. (1,282.54 x 114.3 cm)

WEIGHT: 2,022 lbs (917 kg)

THRUST: 20,800 lbf (92,500 N)

KEY ACCOMPLISHMENT(S): Viking was part of a series of U.S. large-scale, liquid-propellant rockets from a program that began in 1946, designed specifically to replace the diminishing numbers of captured V-2 rockets used for upper-atmospheric research after World War II.

The Viking rockets differed in their configurations and payload capabilities. Reaction Motors, Inc. developed the XLR-10 liquid-propellant rocket engine in Viking 12, launched from White Sands Proving Grounds, New Mexico in 1955, to a height of 144 miles.

SPUTNIK

YEAR: 1957

INVENTOR/MANUFACTURER: OKB-1 ("Experimental Design Bureau")
Korolev Rocket and Space Corporation

COUNTRY OF ORIGIN: U.S.S.R.

OVERALL DIMENSIONS: 23 in. (58 cm)

WEIGHT: 184 lbs (83.6 kg)

LAUNCH VEHICLE: R-7

KEY TRAIT(S): Sputnik's four external radio antennas are its most recognizable feature, designed to look streamlined although it tumbled while in orbit.

KEY ACCOMPLISHMENT(S): On October 4, 1957, the "beep beep" heard 'round the world from Sputnik 1, the first artificial satellite to orbit the Earth, kicked off the Space Race between the U.S.S.R. and the U.S.A.

"Sputnik" is the Russian word for "satellite." Its chief designer, Sergei P. Korolev, had ambitions from the start to make the craft a cultural icon. He insisted on the outer polish, saying "This ball will be exhibited in museums!" However, Sputniks now on display are replicas. Sputnik 1 was entirely incinerated upon re-entry, and all that remains is a metal arming key that prevented contact between the batteries and the radio transmitter. This key, which was removed before launch, is now housed in the Smithsonian.

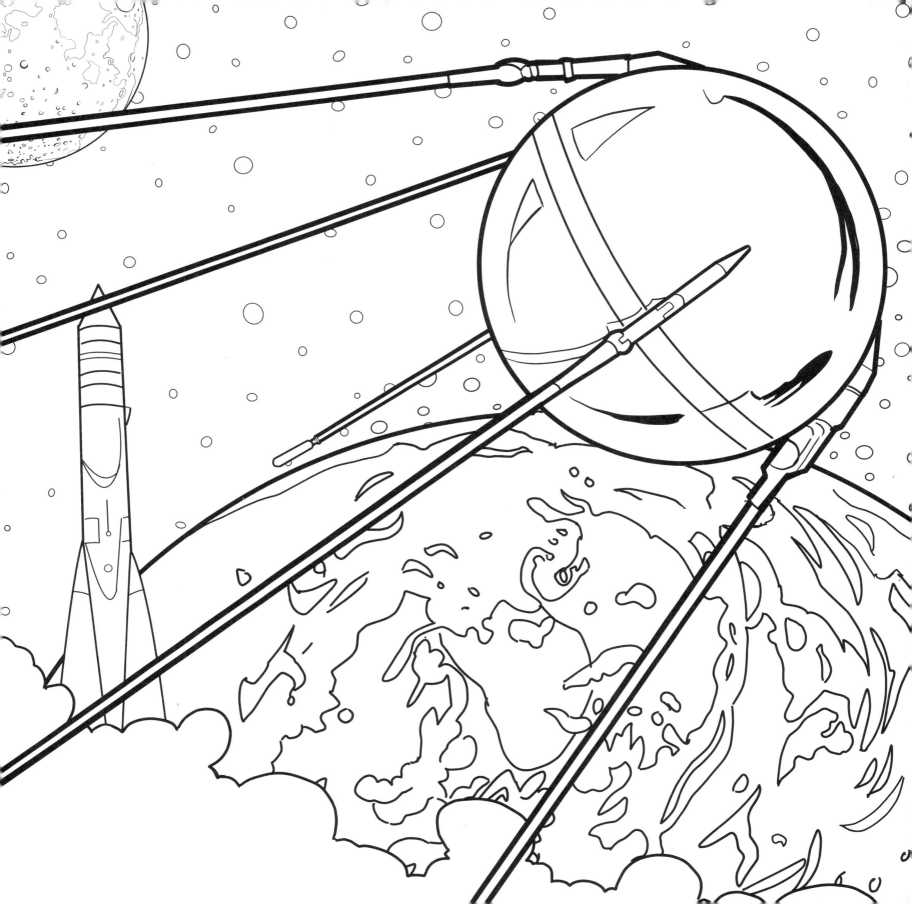

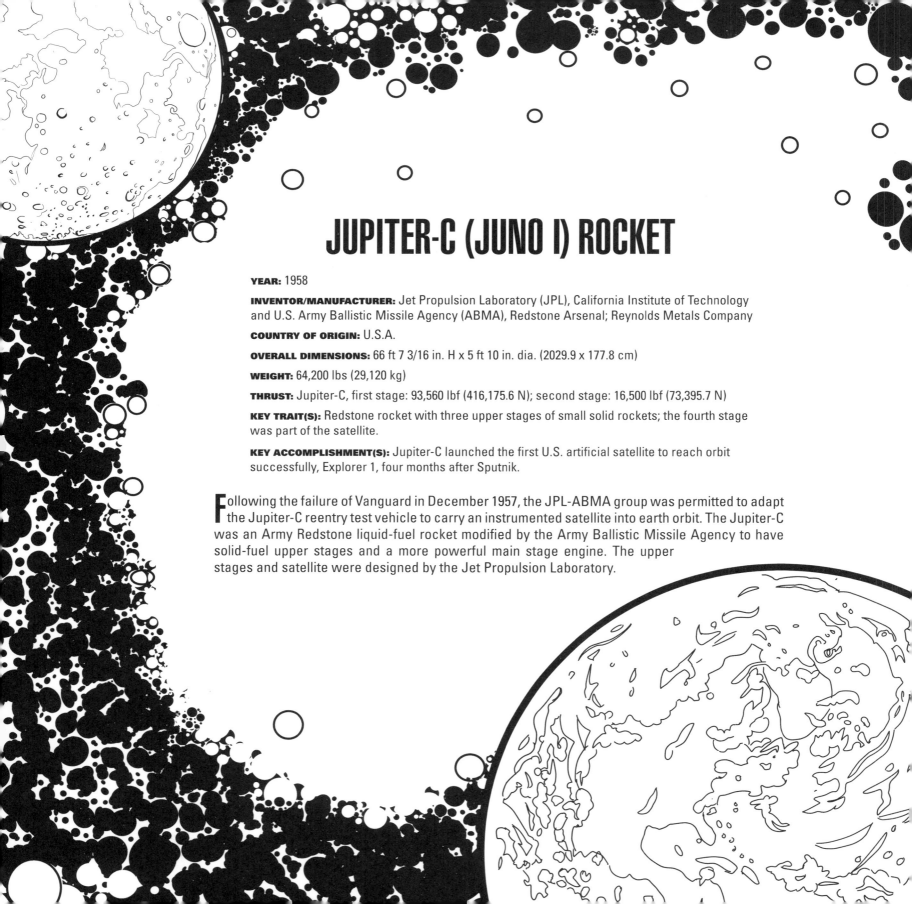

JUPITER-C (JUNO I) ROCKET

YEAR: 1958

INVENTOR/MANUFACTURER: Jet Propulsion Laboratory (JPL), California Institute of Technology and U.S. Army Ballistic Missile Agency (ABMA), Redstone Arsenal; Reynolds Metals Company

COUNTRY OF ORIGIN: U.S.A.

OVERALL DIMENSIONS: 66 ft 7 3/16 in. H x 5 ft 10 in. dia. (2029.9 x 177.8 cm)

WEIGHT: 64,200 lbs (29,120 kg)

THRUST: Jupiter-C, first stage: 93,560 lbf (416,175.6 N); second stage: 16,500 lbf (73,395.7 N)

KEY TRAIT(S): Redstone rocket with three upper stages of small solid rockets; the fourth stage was part of the satellite.

KEY ACCOMPLISHMENT(S): Jupiter-C launched the first U.S. artificial satellite to reach orbit successfully, Explorer 1, four months after Sputnik.

Following the failure of Vanguard in December 1957, the JPL-ABMA group was permitted to adapt the Jupiter-C reentry test vehicle to carry an instrumented satellite into earth orbit. The Jupiter-C was an Army Redstone liquid-fuel rocket modified by the Army Ballistic Missile Agency to have solid-fuel upper stages and a more powerful main stage engine. The upper stages and satellite were designed by the Jet Propulsion Laboratory.

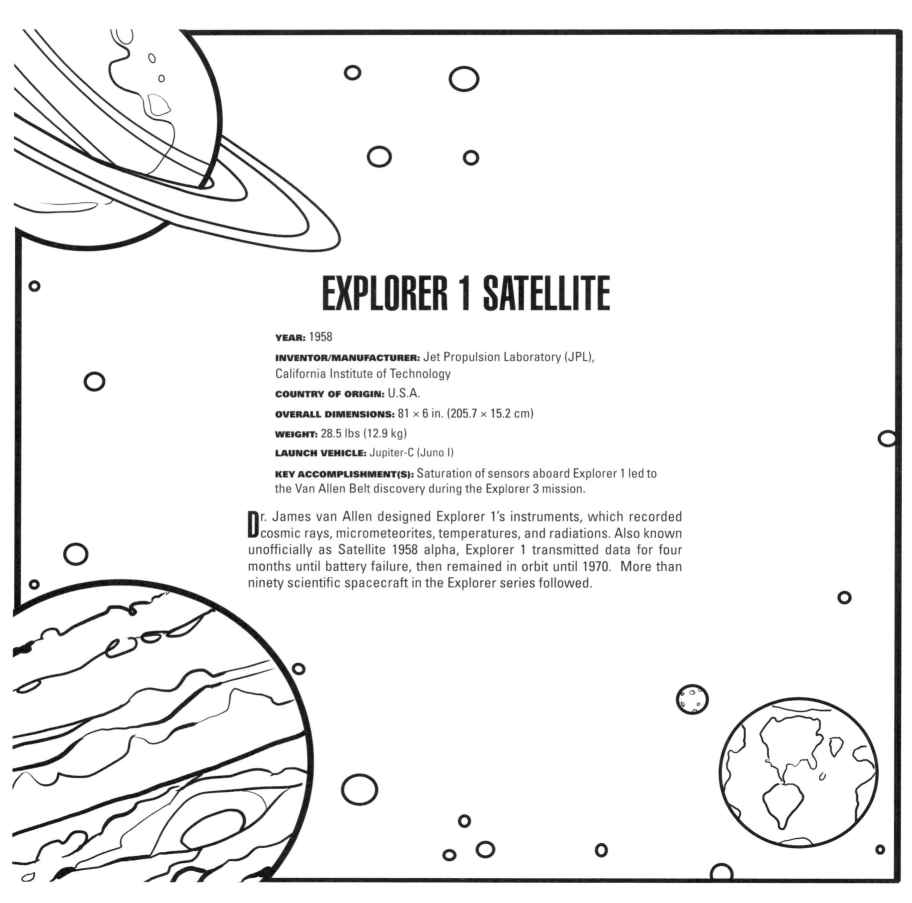

EXPLORER 1 SATELLITE

YEAR: 1958

INVENTOR/MANUFACTURER: Jet Propulsion Laboratory (JPL),
California Institute of Technology

COUNTRY OF ORIGIN: U.S.A.

OVERALL DIMENSIONS: 81 × 6 in. (205.7 × 15.2 cm)

WEIGHT: 28.5 lbs (12.9 kg)

LAUNCH VEHICLE: Jupiter-C (Juno I)

KEY ACCOMPLISHMENT(S): Saturation of sensors aboard Explorer 1 led to
the Van Allen Belt discovery during the Explorer 3 mission.

Dr. James van Allen designed Explorer 1's instruments, which recorded cosmic rays, micrometeorites, temperatures, and radiations. Also known unofficially as Satellite 1958 alpha, Explorer 1 transmitted data for four months until battery failure, then remained in orbit until 1970. More than ninety scientific spacecraft in the Explorer series followed.

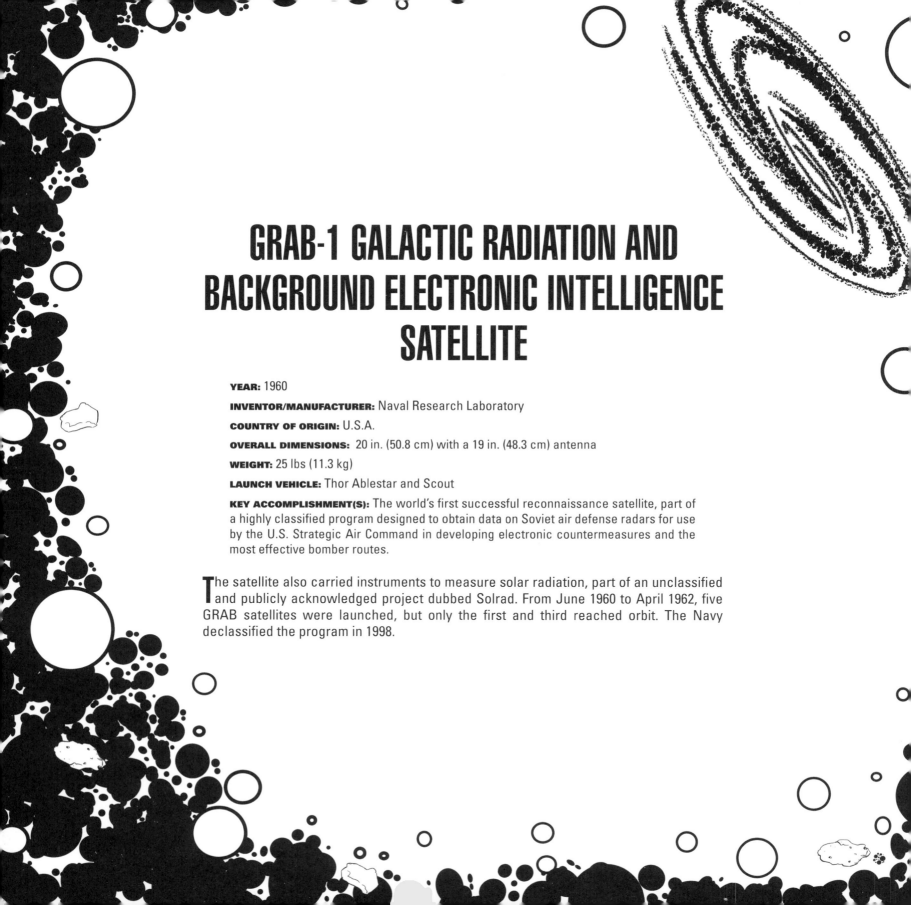

GRAB-1 GALACTIC RADIATION AND BACKGROUND ELECTRONIC INTELLIGENCE SATELLITE

YEAR: 1960

INVENTOR/MANUFACTURER: Naval Research Laboratory

COUNTRY OF ORIGIN: U.S.A.

OVERALL DIMENSIONS: 20 in. (50.8 cm) with a 19 in. (48.3 cm) antenna

WEIGHT: 25 lbs (11.3 kg)

LAUNCH VEHICLE: Thor Ablestar and Scout

KEY ACCOMPLISHMENT(S): The world's first successful reconnaissance satellite, part of a highly classified program designed to obtain data on Soviet air defense radars for use by the U.S. Strategic Air Command in developing electronic countermeasures and the most effective bomber routes.

The satellite also carried instruments to measure solar radiation, part of an unclassified and publicly acknowledged project dubbed Solrad. From June 1960 to April 1962, five GRAB satellites were launched, but only the first and third reached orbit. The Navy declassified the program in 1998.

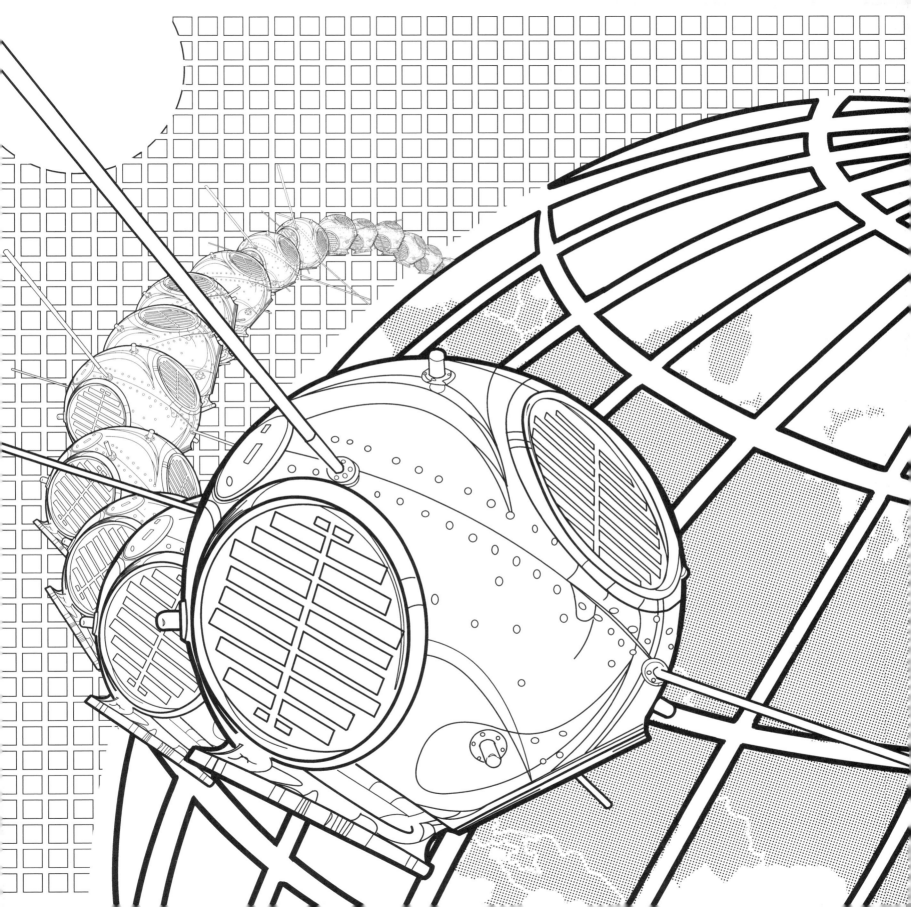

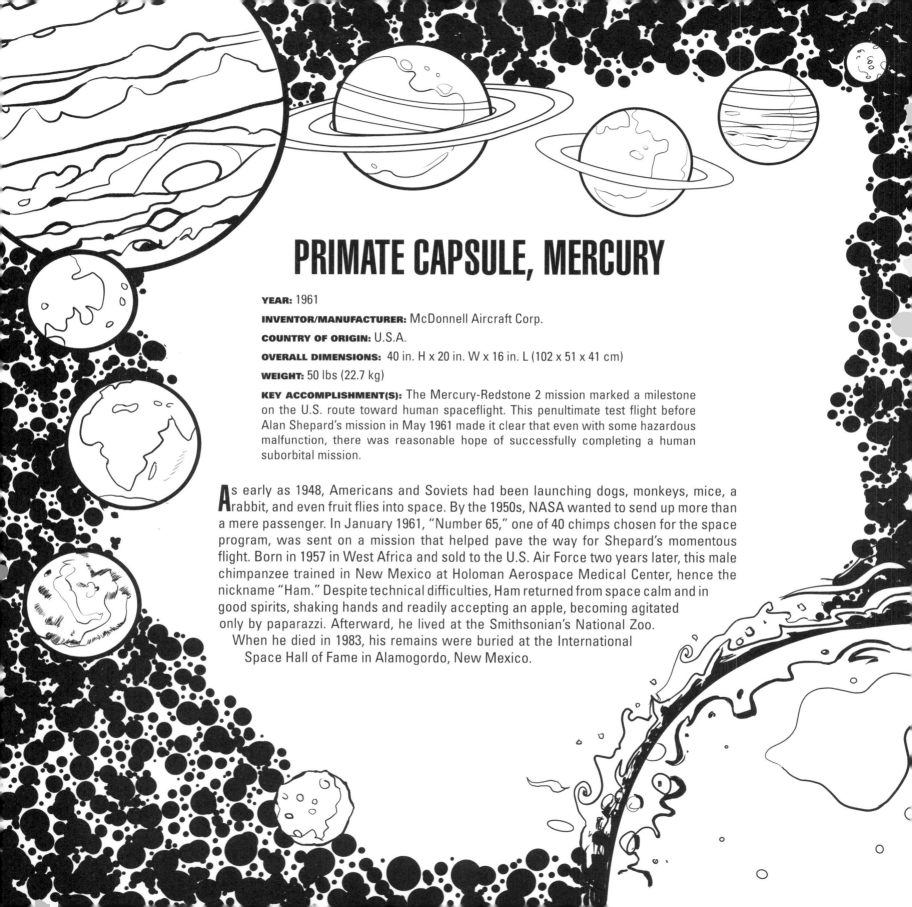

PRIMATE CAPSULE, MERCURY

YEAR: 1961

INVENTOR/MANUFACTURER: McDonnell Aircraft Corp.

COUNTRY OF ORIGIN: U.S.A.

OVERALL DIMENSIONS: 40 in. H x 20 in. W x 16 in. L (102 x 51 x 41 cm)

WEIGHT: 50 lbs (22.7 kg)

KEY ACCOMPLISHMENT(S): The Mercury-Redstone 2 mission marked a milestone on the U.S. route toward human spaceflight. This penultimate test flight before Alan Shepard's mission in May 1961 made it clear that even with some hazardous malfunction, there was reasonable hope of successfully completing a human suborbital mission.

As early as 1948, Americans and Soviets had been launching dogs, monkeys, mice, a rabbit, and even fruit flies into space. By the 1950s, NASA wanted to send up more than a mere passenger. In January 1961, "Number 65," one of 40 chimps chosen for the space program, was sent on a mission that helped pave the way for Shepard's momentous flight. Born in 1957 in West Africa and sold to the U.S. Air Force two years later, this male chimpanzee trained in New Mexico at Holoman Aerospace Medical Center, hence the nickname "Ham." Despite technical difficulties, Ham returned from space calm and in good spirits, shaking hands and readily accepting an apple, becoming agitated only by paparazzi. Afterward, he lived at the Smithsonian's National Zoo. When he died in 1983, his remains were buried at the International Space Hall of Fame in Alamogordo, New Mexico.

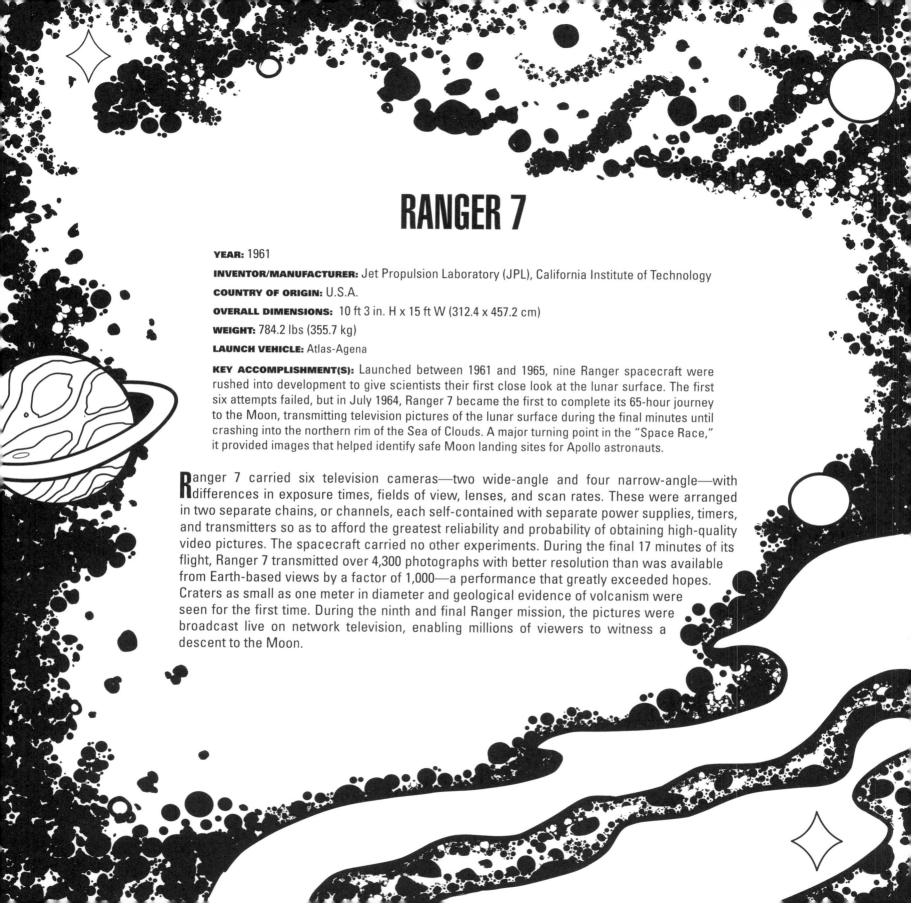

RANGER 7

YEAR: 1961

INVENTOR/MANUFACTURER: Jet Propulsion Laboratory (JPL), California Institute of Technology

COUNTRY OF ORIGIN: U.S.A.

OVERALL DIMENSIONS: 10 ft 3 in. H x 15 ft W (312.4 x 457.2 cm)

WEIGHT: 784.2 lbs (355.7 kg)

LAUNCH VEHICLE: Atlas-Agena

KEY ACCOMPLISHMENT(S): Launched between 1961 and 1965, nine Ranger spacecraft were rushed into development to give scientists their first close look at the lunar surface. The first six attempts failed, but in July 1964, Ranger 7 became the first to complete its 65-hour journey to the Moon, transmitting television pictures of the lunar surface during the final minutes until crashing into the northern rim of the Sea of Clouds. A major turning point in the "Space Race," it provided images that helped identify safe Moon landing sites for Apollo astronauts.

Ranger 7 carried six television cameras—two wide-angle and four narrow-angle—with differences in exposure times, fields of view, lenses, and scan rates. These were arranged in two separate chains, or channels, each self-contained with separate power supplies, timers, and transmitters so as to afford the greatest reliability and probability of obtaining high-quality video pictures. The spacecraft carried no other experiments. During the final 17 minutes of its flight, Ranger 7 transmitted over 4,300 photographs with better resolution than was available from Earth-based views by a factor of 1,000—a performance that greatly exceeded hopes. Craters as small as one meter in diameter and geological evidence of volcanism were seen for the first time. During the ninth and final Ranger mission, the pictures were broadcast live on network television, enabling millions of viewers to witness a descent to the Moon.

TELSTAR

YEAR: 1962

INVENTOR/MANUFACTURER: AT&T

OVERALL DIMENSIONS: 34.5 in. (87.6 cm)

COUNTRY OF ORIGIN: U.S.A.

WEIGHT: 191.5 lbs (86.9 kg)

LAUNCH VEHICLE: Thor-Delta

KEY ACCOMPLISHMENT(S): The world's first active communications satellite, Telstar 1 inaugurated an era of live international television, relaying images between the U.S., France and England upon launch in July 1962.

Telstar 1 received microwave radio signals from ground stations and retransmitted them across vast distances back to Earth. Used to test basic features of communications via space, it became the model for all subsequent communications satellites. Telstar raised an important policy question: Should communications satellites be operated and controlled by private corporations or under government auspices? The United States chose government direction and created two new institutions, COMSAT and INTELSAT, to develop satellite communications, an arrangement that lasted over two decades. In November 1962, prolonged exposure to Van Allen Belt radiation compromised Telstar's electronics; the satellite deactivated in February 1963.

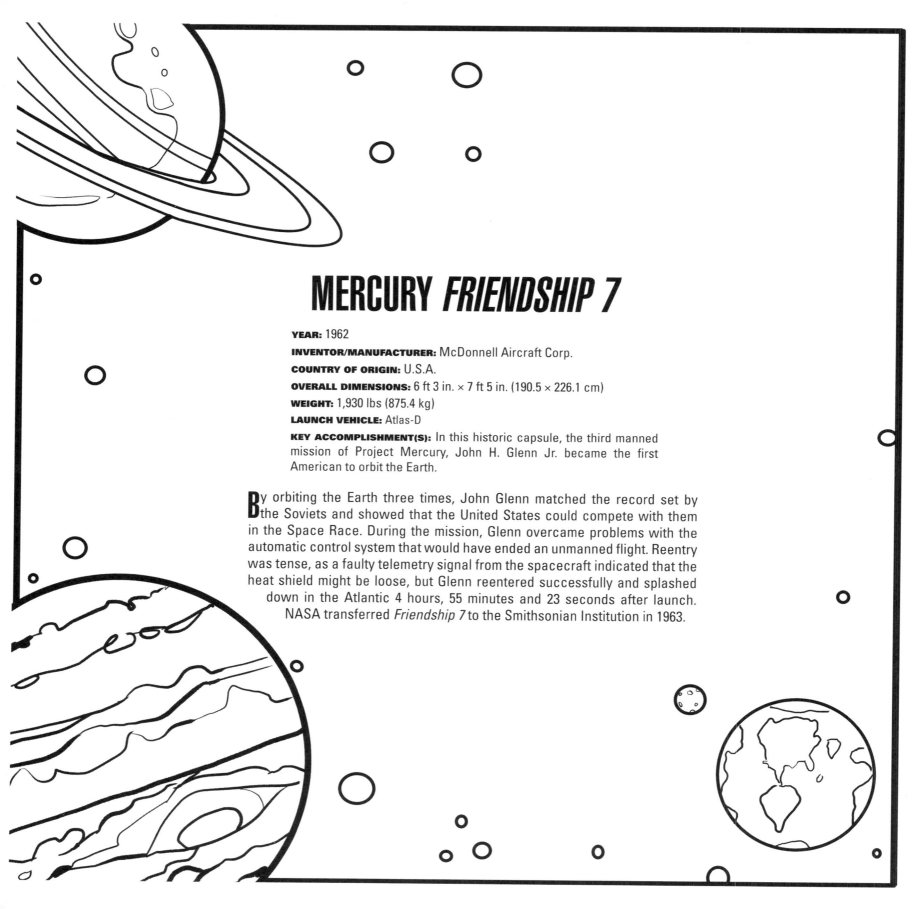

MERCURY *FRIENDSHIP 7*

YEAR: 1962

INVENTOR/MANUFACTURER: McDonnell Aircraft Corp.

COUNTRY OF ORIGIN: U.S.A.

OVERALL DIMENSIONS: 6 ft 3 in. × 7 ft 5 in. (190.5 × 226.1 cm)

WEIGHT: 1,930 lbs (875.4 kg)

LAUNCH VEHICLE: Atlas-D

KEY ACCOMPLISHMENT(S): In this historic capsule, the third manned mission of Project Mercury, John H. Glenn Jr. became the first American to orbit the Earth.

By orbiting the Earth three times, John Glenn matched the record set by the Soviets and showed that the United States could compete with them in the Space Race. During the mission, Glenn overcame problems with the automatic control system that would have ended an unmanned flight. Reentry was tense, as a faulty telemetry signal from the spacecraft indicated that the heat shield might be loose, but Glenn reentered successfully and splashed down in the Atlantic 4 hours, 55 minutes and 23 seconds after launch. NASA transferred *Friendship 7* to the Smithsonian Institution in 1963.

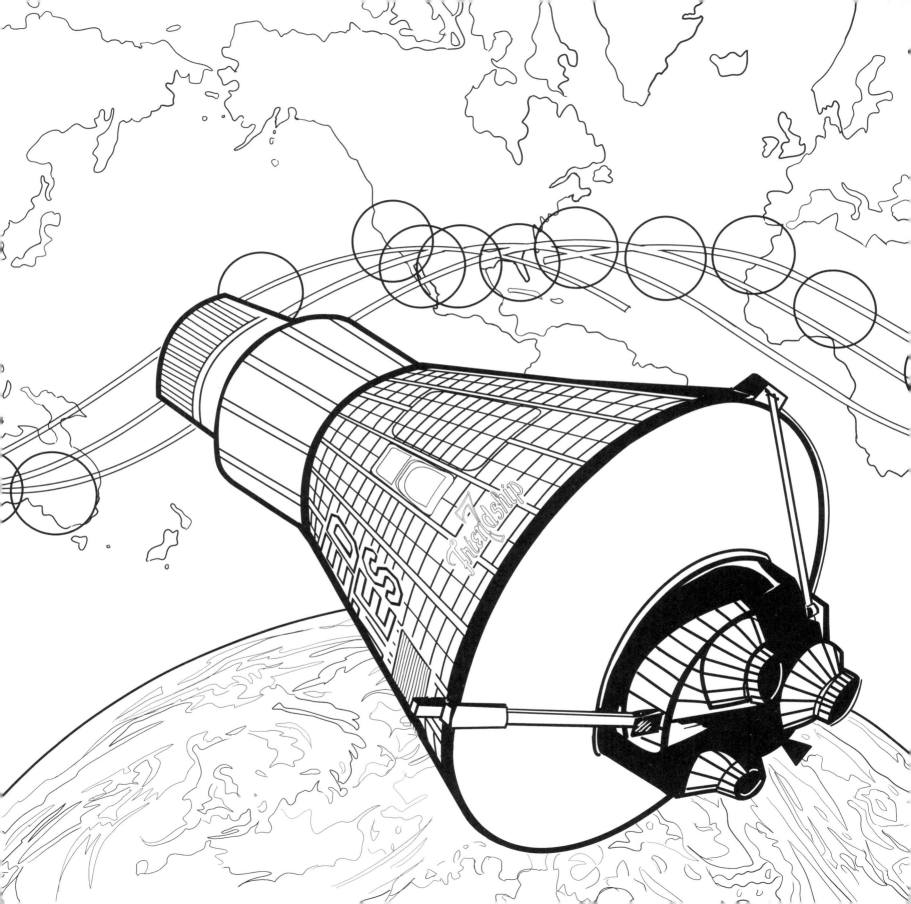

GEMINI IV

YEAR: 1965

INVENTOR/MANUFACTURER: McDonnell Aircraft Corp.

COUNTRY OF ORIGIN: U.S.A.

OVERALL DIMENSIONS: 7 ft 6 in. W x 8 ft 3 in. H (228.6 × 251.5 cm)

WEIGHT: 3,020 lbs (1,369.9 kg)

LAUNCH VEHICLE: Titan II

KEY TRAITS: Bell-shaped, two-man capsule with two windows and two crew egress hatches

KEY ACCOMPLISHMENTS: Astronaut Ed White ventured from the Gemini IV early in the mission to achieve the first American spacewalk, a major step toward living and working in space, and a critical bridge between the one-person Mercury program and the more sophisticated Apollo Moon missions.

Gemini IV's June 1965 flight lasted four days and included a live audio NASA broadcast of Ed White's 22-minute "extravehicular activity" (EVA). Ten weeks earlier, Soviet cosmonaut Alexei Leonov had become the first human to "walk in space." Other experiments during this flight included Earth photography, space radiation measurements, and the medical effects of prolonged weightlessness.

LITTON PRESSURE SUIT RX-2

YEAR: 1965

INVENTOR/MANUFACTURER: Litton Guidance and Control Systems

COUNTRY OF ORIGIN: U.S.A.

OVERALL DIMENSIONS: 5 ft 9 in. L x 2 ft 5 in. W x 1 ft 1 in. H (175.3 x 73.7 x 33 cm)

WEIGHT: 3.2 lbs (1.47 kg)

KEY ACCOMPLISHMENTS: Designed to maintain an almost perfectly constant volume while enabling a full range of body motions, these suits could operate at higher pressure, thus reducing the time-consuming oxygen pre-breathing period before extravehicular activities (EVA).

Litton began adapting its line of vacuum chamber suits first for the Air Force in 1955 and then for NASA in 1964, as a solution to the problems of bending at the hips in previous models. The RX-2 and RX-2A models differed in the weight and design of the upper body only; thus, the lower sections were interchangeable. This was the first of the constant-volume suits to have all-metal rolling convolute. This suit is constructed predominantly of aluminum and has electrical connectors and fittings. Due to bulk considerations, the constant-volume suits were never used as mission equipment.

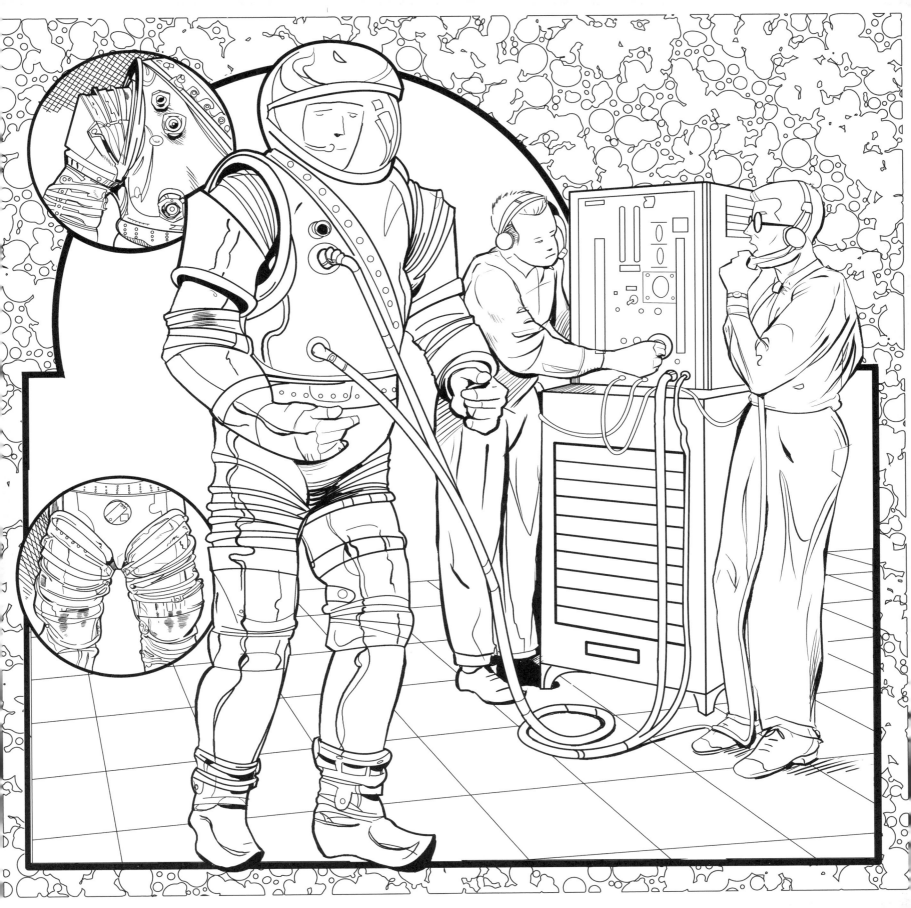

SURVEYOR LUNAR LANDER

YEAR: 1966-1968

INVENTOR/MANUFACTURER: Hughes Aircraft Co.

COUNTRY OF ORIGIN: U.S.A.

OVERALL DIMENSIONS: 10 ft H x 14 ft W (304.8 x 426.7 cm)

WEIGHT: 649 lbs (294.3 kg)

LAUNCH VEHICLE: Atlas-Centaur

KEY ACCOMPLISHMENTS: The first U.S. probes to soft-land on the Moon, Surveyor was designed to provide data about its surface, and possible atmosphere, taking detailed pictures with an onboard TV camera. Seven Surveyors launched starting in May 1966, and all but two were successful, returning over 88,000 high resolution photographs and invaluable detailed data on the nature and strength of the lunar surface. The program also supplied valuable experience controlling spacecraft far from Earth and near another celestial body, as well as the remote manipulation of instruments in space.

Unlike Ranger, which crash landed on the lunar surface, and Lunar Orbiter, which operated from lunar orbit, Surveyor was designed to survive its trip, land safely, and operate for extended periods of time. Surveyor 3 had been on the Moon for 2½ years when the Apollo 12 crew arrived in 1969. Astronauts Charles Conrad Jr. and Alan Bean removed its television camera, surface sampler and some tubing and brought them back to Earth for analysis.

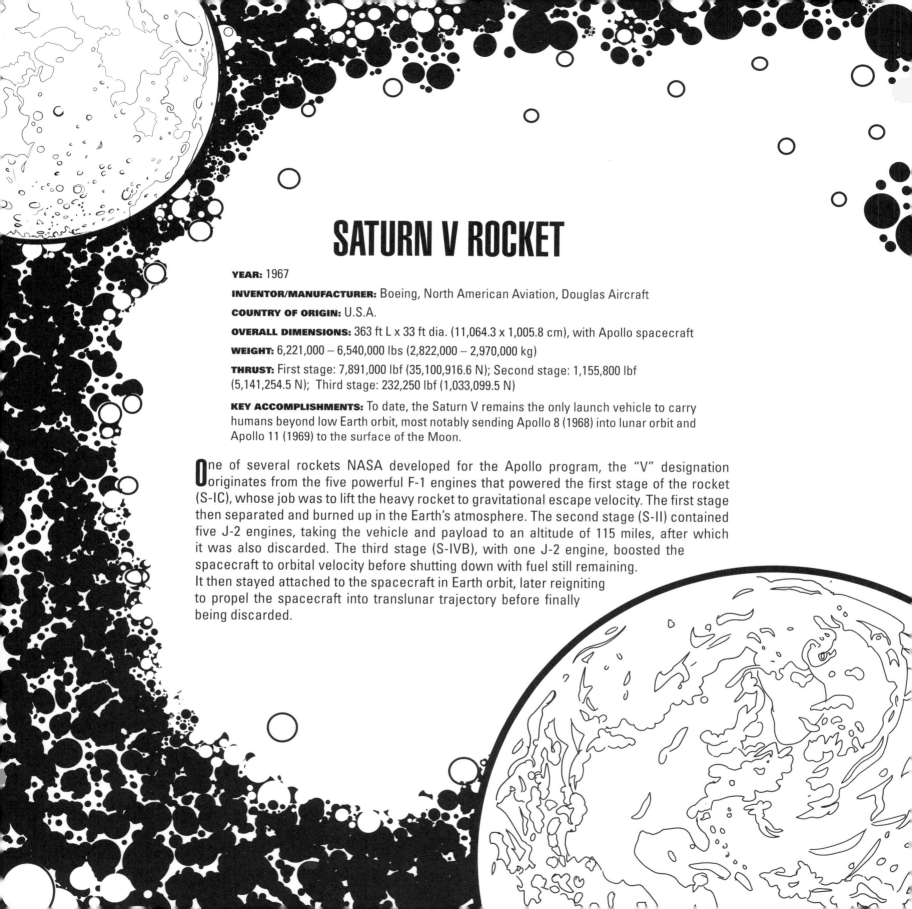

SATURN V ROCKET

YEAR: 1967

INVENTOR/MANUFACTURER: Boeing, North American Aviation, Douglas Aircraft

COUNTRY OF ORIGIN: U.S.A.

OVERALL DIMENSIONS: 363 ft L x 33 ft dia. (11,064.3 x 1,005.8 cm), with Apollo spacecraft

WEIGHT: 6,221,000 – 6,540,000 lbs (2,822,000 – 2,970,000 kg)

THRUST: First stage: 7,891,000 lbf (35,100,916.6 N); Second stage: 1,155,800 lbf (5,141,254.5 N); Third stage: 232,250 lbf (1,033,099.5 N)

KEY ACCOMPLISHMENTS: To date, the Saturn V remains the only launch vehicle to carry humans beyond low Earth orbit, most notably sending Apollo 8 (1968) into lunar orbit and Apollo 11 (1969) to the surface of the Moon.

One of several rockets NASA developed for the Apollo program, the "V" designation originates from the five powerful F-1 engines that powered the first stage of the rocket (S-IC), whose job was to lift the heavy rocket to gravitational escape velocity. The first stage then separated and burned up in the Earth's atmosphere. The second stage (S-II) contained five J-2 engines, taking the vehicle and payload to an altitude of 115 miles, after which it was also discarded. The third stage (S-IVB), with one J-2 engine, boosted the spacecraft to orbital velocity before shutting down with fuel still remaining. It then stayed attached to the spacecraft in Earth orbit, later reigniting to propel the spacecraft into translunar trajectory before finally being discarded.

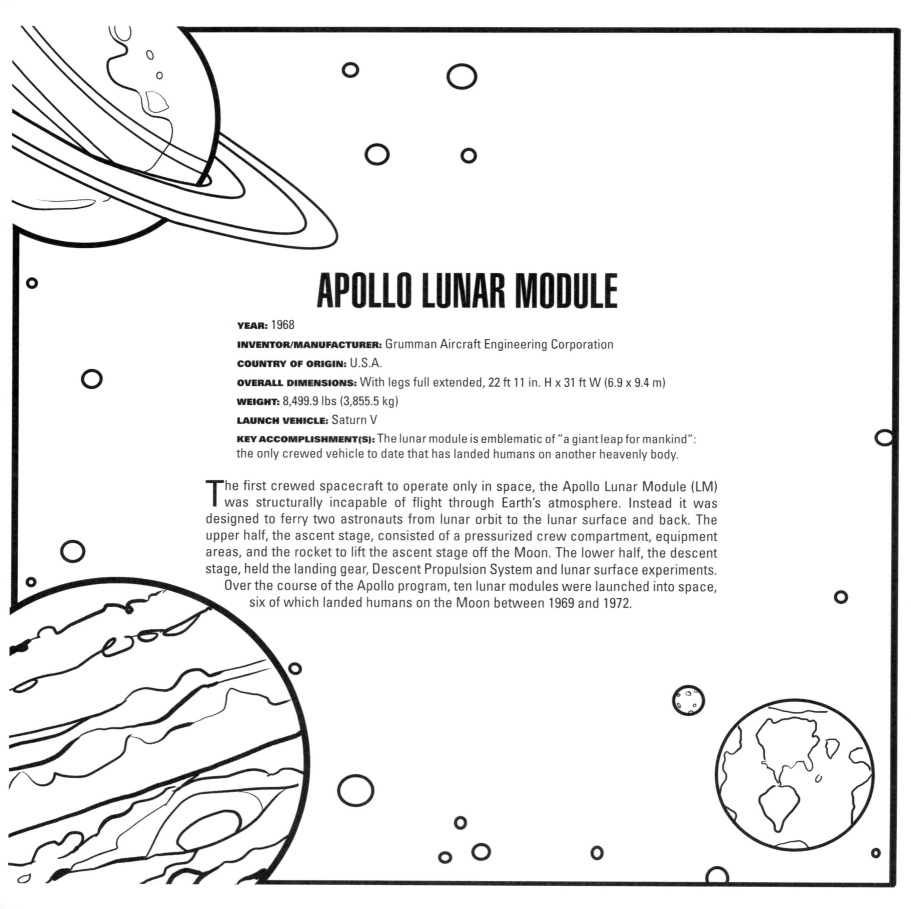

APOLLO LUNAR MODULE

YEAR: 1968

INVENTOR/MANUFACTURER: Grumman Aircraft Engineering Corporation

COUNTRY OF ORIGIN: U.S.A.

OVERALL DIMENSIONS: With legs full extended, 22 ft 11 in. H x 31 ft W (6.9 x 9.4 m)

WEIGHT: 8,499.9 lbs (3,855.5 kg)

LAUNCH VEHICLE: Saturn V

KEY ACCOMPLISHMENT(S): The lunar module is emblematic of "a giant leap for mankind": the only crewed vehicle to date that has landed humans on another heavenly body.

The first crewed spacecraft to operate only in space, the Apollo Lunar Module (LM) was structurally incapable of flight through Earth's atmosphere. Instead it was designed to ferry two astronauts from lunar orbit to the lunar surface and back. The upper half, the ascent stage, consisted of a pressurized crew compartment, equipment areas, and the rocket to lift the ascent stage off the Moon. The lower half, the descent stage, held the landing gear, Descent Propulsion System and lunar surface experiments.

Over the course of the Apollo program, ten lunar modules were launched into space, six of which landed humans on the Moon between 1969 and 1972.

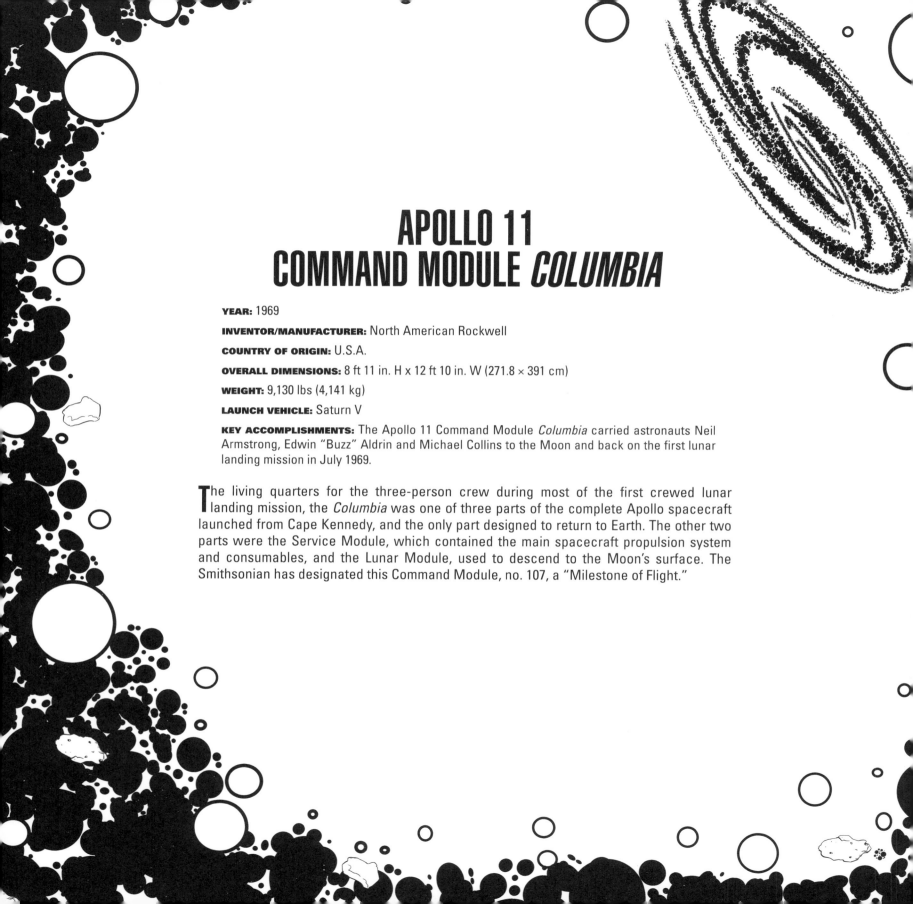

APOLLO 11
COMMAND MODULE *COLUMBIA*

YEAR: 1969

INVENTOR/MANUFACTURER: North American Rockwell

COUNTRY OF ORIGIN: U.S.A.

OVERALL DIMENSIONS: 8 ft 11 in. H x 12 ft 10 in. W (271.8 × 391 cm)

WEIGHT: 9,130 lbs (4,141 kg)

LAUNCH VEHICLE: Saturn V

KEY ACCOMPLISHMENTS: The Apollo 11 Command Module *Columbia* carried astronauts Neil Armstrong, Edwin "Buzz" Aldrin and Michael Collins to the Moon and back on the first lunar landing mission in July 1969.

The living quarters for the three-person crew during most of the first crewed lunar landing mission, the *Columbia* was one of three parts of the complete Apollo spacecraft launched from Cape Kennedy, and the only part designed to return to Earth. The other two parts were the Service Module, which contained the main spacecraft propulsion system and consumables, and the Lunar Module, used to descend to the Moon's surface. The Smithsonian has designated this Command Module, no. 107, a "Milestone of Flight."

PRESSURE SUIT, A7-L, ARMSTRONG, APOLLO 11, FLOWN

YEAR: 1969

INVENTOR/MANUFACTURER: ILC Industries, Inc.

COUNTRY OF ORIGIN: U.S.A.

OVERALL DIMENSIONS: 6 ft 9/16 in. L x 2 ft 8 5/16 in. W x 11 in. H (184 x 82 x 28 cm)

WEIGHT: 80 lbs (36.3 kg)

KEY ACCOMPLISHMENT(S): This spacesuit was worn by astronaut Neil Armstrong, Commander of the Apollo 11 mission, when he took the historic first steps on the Moon.

These spacesuits were designed to provide life-sustaining environments for astronauts during periods of extravehicular activity (EVA) or unpressurized spacecraft operation. They permitted maximum mobility and were designed to be worn with relative comfort for up to 115 hours in conjunction with the liquid cooling garment or if needed, up to 14 days in an unpressurized mode.

LUNAR ROVING VEHICLE

YEAR: 1971

INVENTOR/MANUFACTURER: Boeing

COUNTRY OF ORIGIN: U.S.A.

OVERALL DIMENSIONS: 4 ft H x 5 ft W x 8 ft L (121.9 × 152.4 × 243.8 cm)

WEIGHT: 537 lbs (243.6 kg)

KEY ACCOMPLISHMENT(S): Apollo 15 marked the first excursion of the Lunar Roving Vehicle (LRV), a "dune buggy" that accompanied the astronauts on the last three Apollo lunar landings.

Stowed in the descent stage of the Lunar Module and deployed upon arrival at the lunar surface, the LRV was operated with a spacecraft "stick," rather than a steering wheel, and could move forward and backwards. In addition to the flight vehicles, Boeing manufactured eight non-flight units for development and testing. One of these, the "Qualification Test Unit," a very close replica of the actual mission unit, was deliberately subjected in special test chambers to conditions many times more severe than expected. As test stresses had left the unit unsuitable for safe use in space, NASA transferred it to the Smithsonian in 1975.

PIONEER 10/11

YEAR: 1972

INVENTOR/MANUFACTURER: TRW, Inc.

COUNTRY OF ORIGIN: U.S.A.

OVERALL DIMENSIONS: 9 ft W x 9 ft 6 in. L x 9 ft dia. (274.3 x 289.5 x 274.3 cm)

WEIGHT: 568 lbs (257.6 kg)

LAUNCH VEHICLES: Atlas-Centaur

KEY ACCOMPLISHMENT(S): The Pioneer probes, launched in 1972 and 1973, were the first spacecraft to traverse the asteroid belt and fly by Jupiter and Saturn. Now on their way out of the solar system to explore the galaxy, Pioneer 10 and 11 excited people about deep space exploration. Pioneer 10 carries a plaque designed to inform any potential intelligent life about the spacecraft and its origins.

For over 30 years, the Pioneer 10 spacecraft sent photographs and scientific information back to Earth on such phenomena as solar wind and cosmic rays. On its flight to Jupiter, it reached speeds of 32,400 mph (52,100 kph), making it one of the fastest human-made objects. Pioneer's last, very weak signal was received on January 22, 2003, from approximately 7.6 billion miles (12.2 billion km) from Earth. Now further into space than any other object sent from Earth, Pioneer 10's radioisotope power source has degraded, says NASA, and is not likely to allow future transmissions.

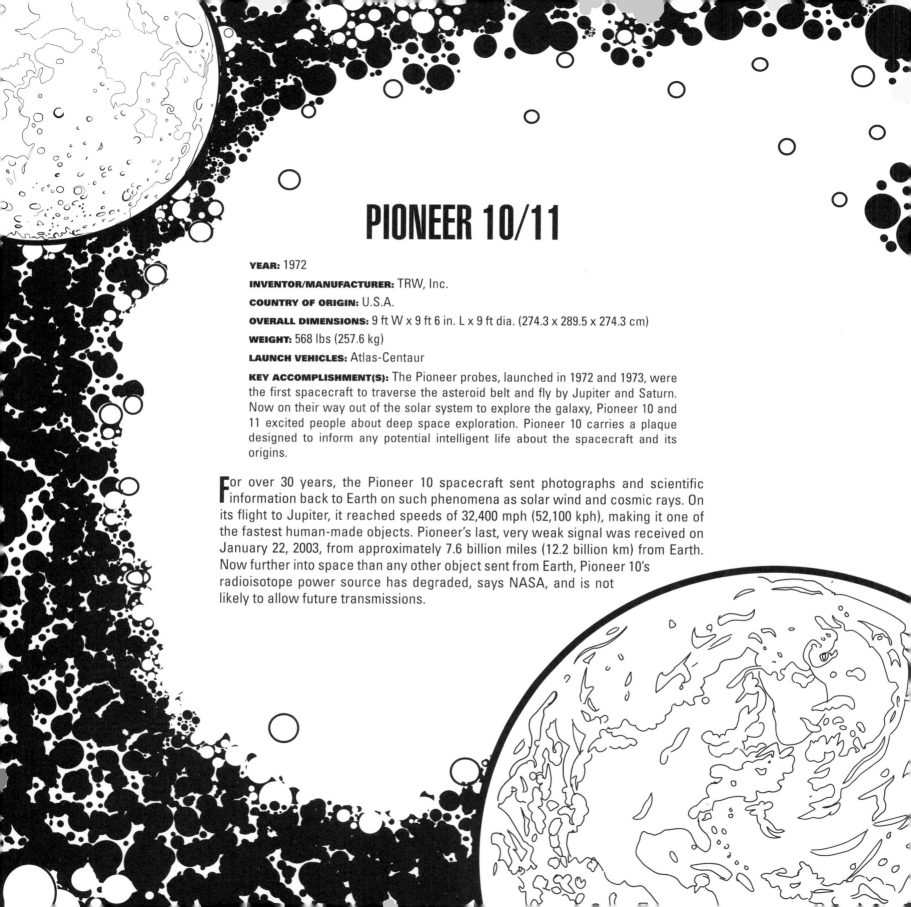

PRESSURE SUIT, SOKOL KV-2, DENNIS TITO

YEAR: 1973

INVENTOR/MANUFACTURER: Zvezda

COUNTRY OF ORIGIN: U.S.S.R.

OVERALL DIMENSIONS: 5 ft 3 1/2 in. L x 2 ft 4 in. W x 9 in. H (161.3 x 71 x 22.9 cm)

WEIGHT: 22 lbs (10 kg)

KEY ACCOMPLISHMENT(S): Introduced in 1973 and still in use in 2020, the Soviet Sokol space suit (Russian for "falcon"), was worn by all Soyuz spacecraft cosmonauts and, most recently, by California businessman Dennis Tito when he became the first tourist in space on April 28, 2001.

Described by its makers as a rescue suit, the Sokol KV-2 was designed in the early 1970s to protect cosmonauts during launch, landing and emergencies. The plugs and tubes extending from the suit connect to life-support systems built into the Soyuz spacecraft. The suit cannot be used outside the spacecraft in a spacewalk or extravehicular activity (EVA) but is meant instead to keep the wearer alive in case of accidental depressurization in the spacecraft.

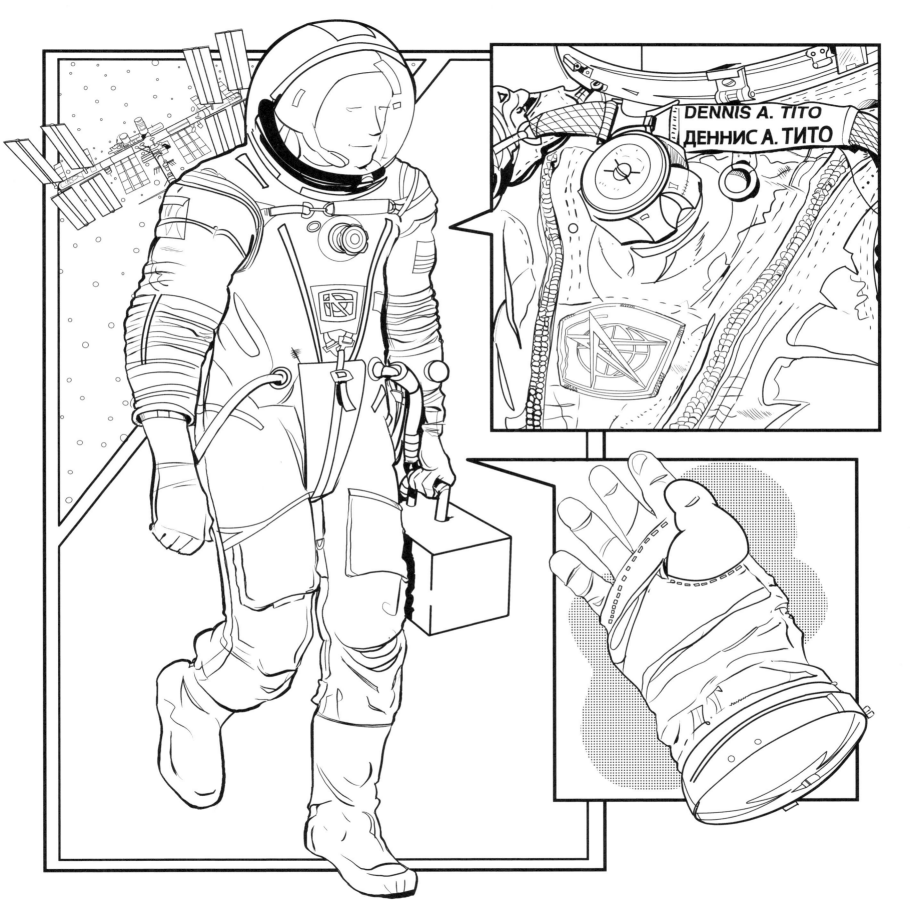

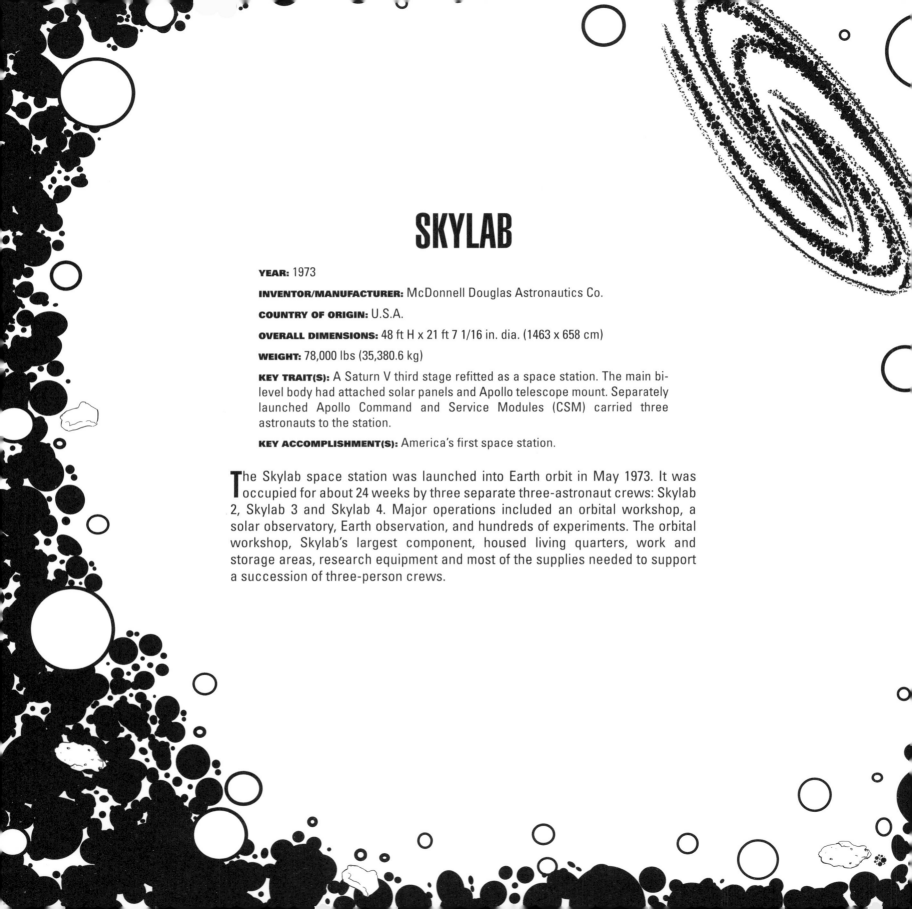

SKYLAB

YEAR: 1973

INVENTOR/MANUFACTURER: McDonnell Douglas Astronautics Co.

COUNTRY OF ORIGIN: U.S.A.

OVERALL DIMENSIONS: 48 ft H x 21 ft 7 1/16 in. dia. (1463 x 658 cm)

WEIGHT: 78,000 lbs (35,380.6 kg)

KEY TRAIT(S): A Saturn V third stage refitted as a space station. The main bi-level body had attached solar panels and Apollo telescope mount. Separately launched Apollo Command and Service Modules (CSM) carried three astronauts to the station.

KEY ACCOMPLISHMENT(S): America's first space station.

The Skylab space station was launched into Earth orbit in May 1973. It was occupied for about 24 weeks by three separate three-astronaut crews: Skylab 2, Skylab 3 and Skylab 4. Major operations included an orbital workshop, a solar observatory, Earth observation, and hundreds of experiments. The orbital workshop, Skylab's largest component, housed living quarters, work and storage areas, research equipment and most of the supplies needed to support a succession of three-person crews.

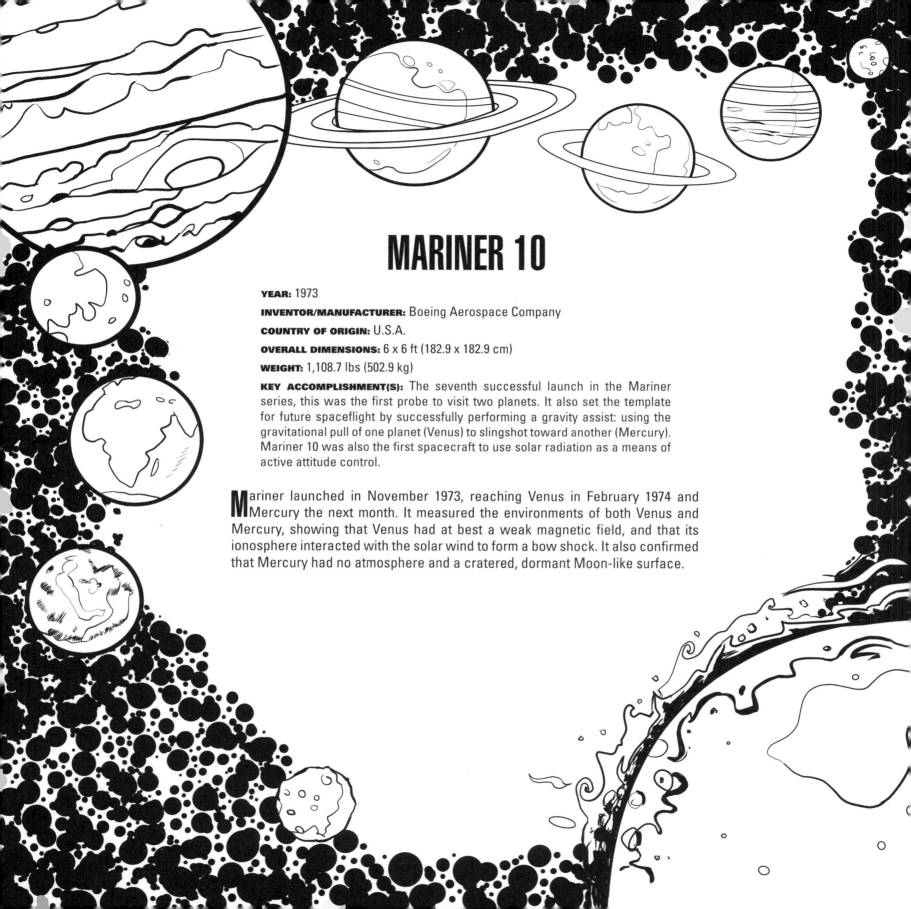

MARINER 10

YEAR: 1973

INVENTOR/MANUFACTURER: Boeing Aerospace Company

COUNTRY OF ORIGIN: U.S.A.

OVERALL DIMENSIONS: 6 x 6 ft (182.9 x 182.9 cm)

WEIGHT: 1,108.7 lbs (502.9 kg)

KEY ACCOMPLISHMENT(S): The seventh successful launch in the Mariner series, this was the first probe to visit two planets. It also set the template for future spaceflight by successfully performing a gravity assist: using the gravitational pull of one planet (Venus) to slingshot toward another (Mercury). Mariner 10 was also the first spacecraft to use solar radiation as a means of active attitude control.

Mariner launched in November 1973, reaching Venus in February 1974 and Mercury the next month. It measured the environments of both Venus and Mercury, showing that Venus had at best a weak magnetic field, and that its ionosphere interacted with the solar wind to form a bow shock. It also confirmed that Mercury had no atmosphere and a cratered, dormant Moon-like surface.

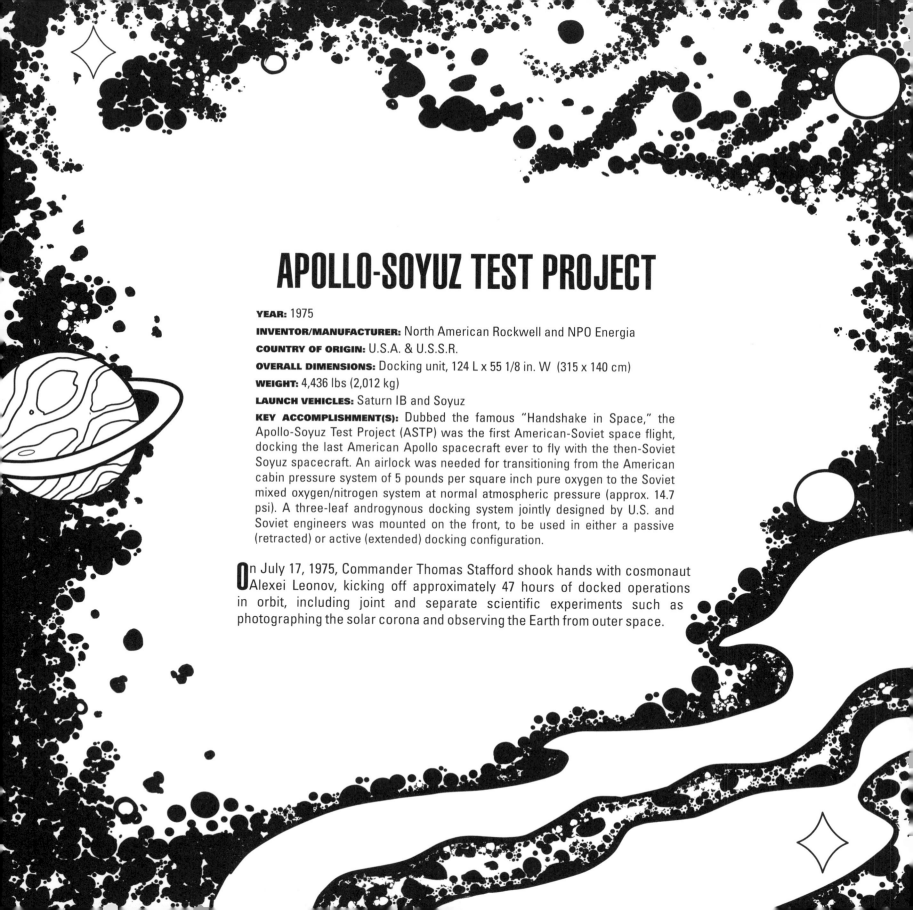

APOLLO-SOYUZ TEST PROJECT

YEAR: 1975

INVENTOR/MANUFACTURER: North American Rockwell and NPO Energia

COUNTRY OF ORIGIN: U.S.A. & U.S.S.R.

OVERALL DIMENSIONS: Docking unit, 124 L x 55 1/8 in. W (315 x 140 cm)

WEIGHT: 4,436 lbs (2,012 kg)

LAUNCH VEHICLES: Saturn IB and Soyuz

KEY ACCOMPLISHMENT(S): Dubbed the famous "Handshake in Space," the Apollo-Soyuz Test Project (ASTP) was the first American-Soviet space flight, docking the last American Apollo spacecraft ever to fly with the then-Soviet Soyuz spacecraft. An airlock was needed for transitioning from the American cabin pressure system of 5 pounds per square inch pure oxygen to the Soviet mixed oxygen/nitrogen system at normal atmospheric pressure (approx. 14.7 psi). A three-leaf androgynous docking system jointly designed by U.S. and Soviet engineers was mounted on the front, to be used in either a passive (retracted) or active (extended) docking configuration.

On July 17, 1975, Commander Thomas Stafford shook hands with cosmonaut Alexei Leonov, kicking off approximately 47 hours of docked operations in orbit, including joint and separate scientific experiments such as photographing the solar corona and observing the Earth from outer space.

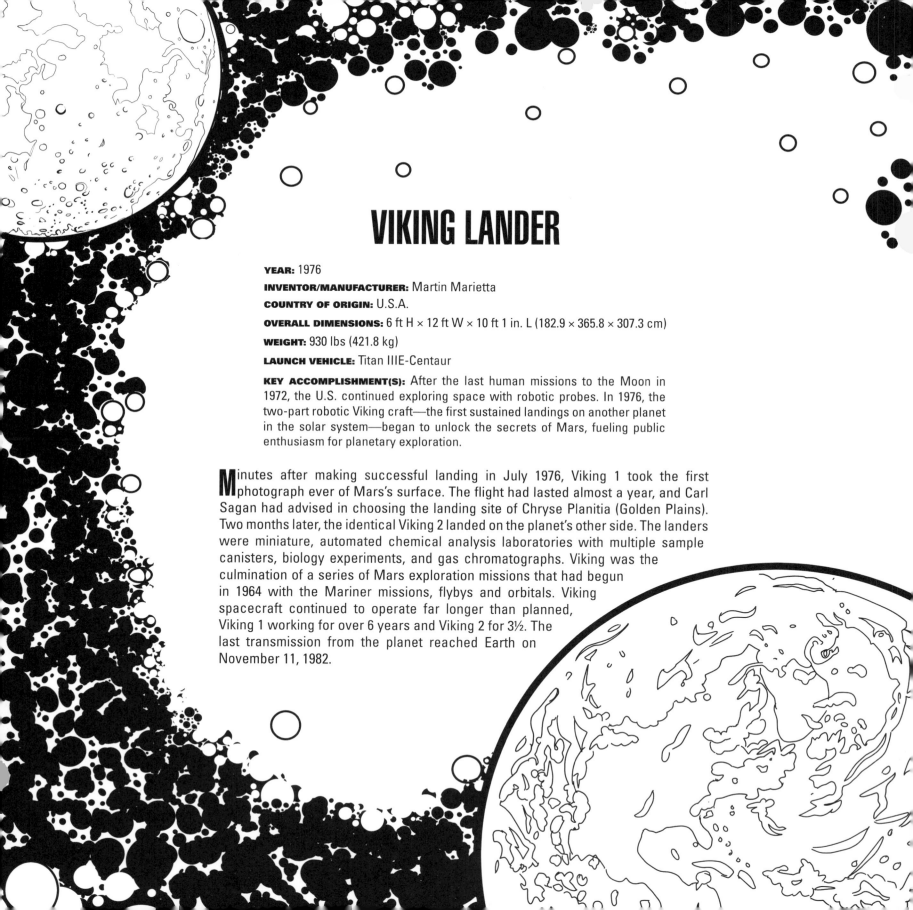

VIKING LANDER

YEAR: 1976

INVENTOR/MANUFACTURER: Martin Marietta

COUNTRY OF ORIGIN: U.S.A.

OVERALL DIMENSIONS: 6 ft H × 12 ft W × 10 ft 1 in. L (182.9 × 365.8 × 307.3 cm)

WEIGHT: 930 lbs (421.8 kg)

LAUNCH VEHICLE: Titan IIIE-Centaur

KEY ACCOMPLISHMENT(S): After the last human missions to the Moon in 1972, the U.S. continued exploring space with robotic probes. In 1976, the two-part robotic Viking craft—the first sustained landings on another planet in the solar system—began to unlock the secrets of Mars, fueling public enthusiasm for planetary exploration.

Minutes after making successful landing in July 1976, Viking 1 took the first photograph ever of Mars's surface. The flight had lasted almost a year, and Carl Sagan had advised in choosing the landing site of Chryse Planitia (Golden Plains). Two months later, the identical Viking 2 landed on the planet's other side. The landers were miniature, automated chemical analysis laboratories with multiple sample canisters, biology experiments, and gas chromatographs. Viking was the culmination of a series of Mars exploration missions that had begun in 1964 with the Mariner missions, flybys and orbitals. Viking spacecraft continued to operate far longer than planned, Viking 1 working for over 6 years and Viking 2 for 3½. The last transmission from the planet reached Earth on November 11, 1982.

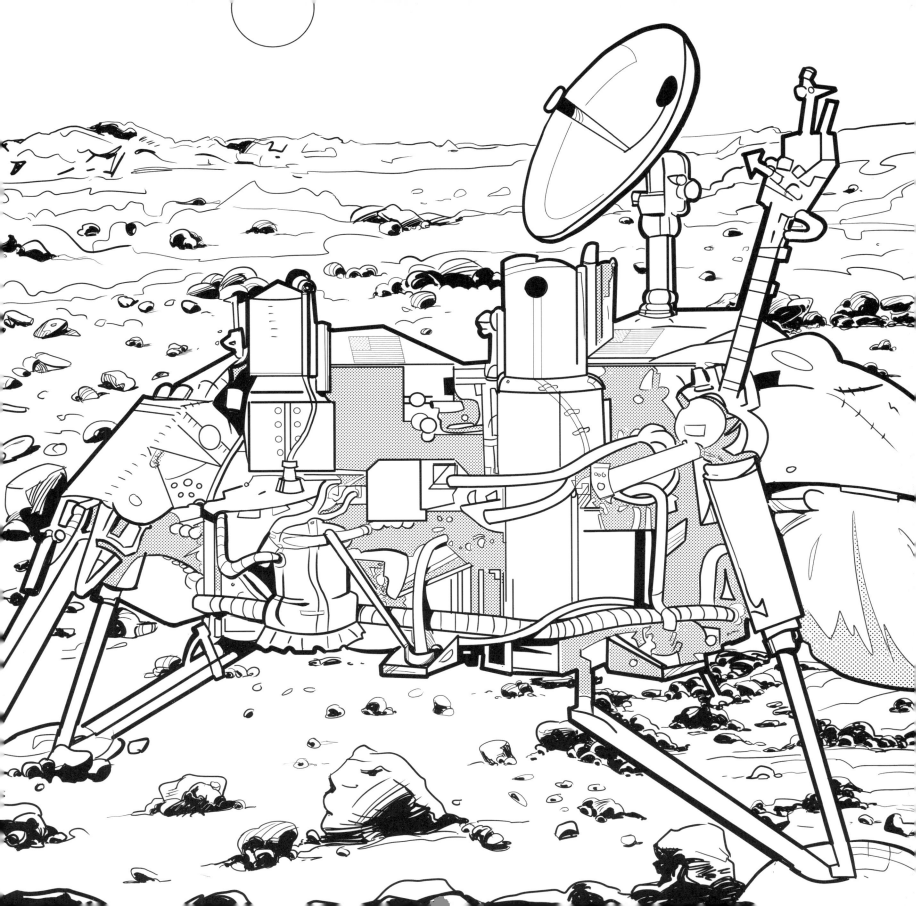

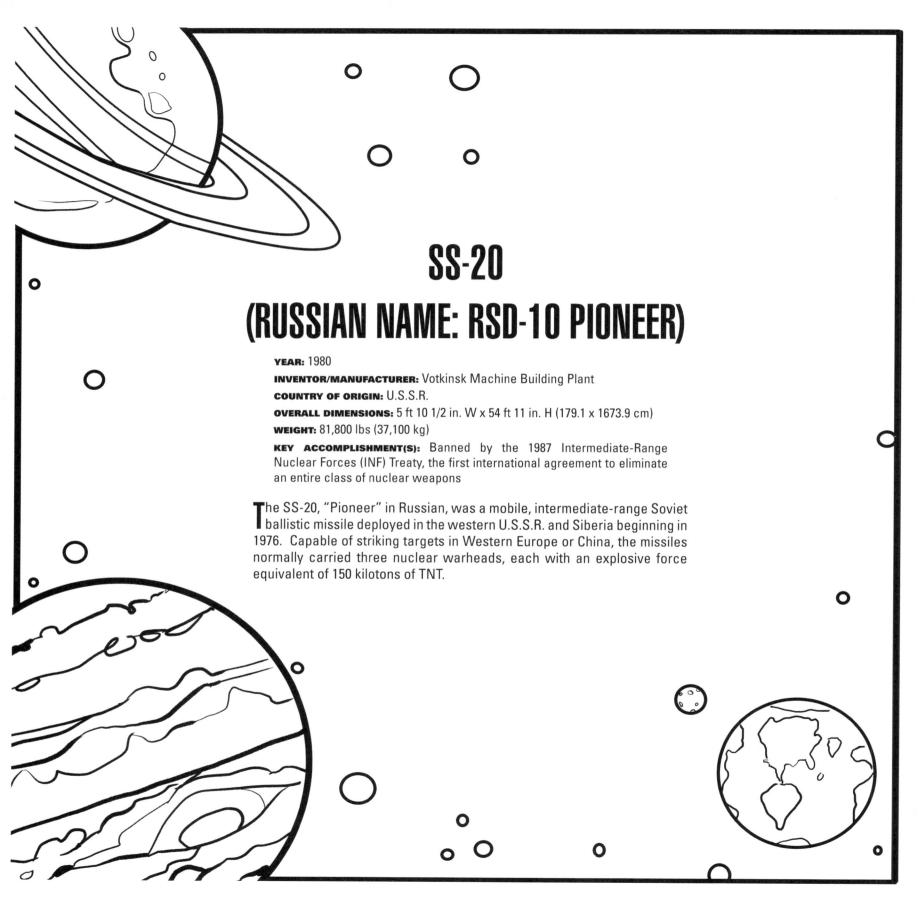

SS-20
(RUSSIAN NAME: RSD-10 PIONEER)

YEAR: 1980

INVENTOR/MANUFACTURER: Votkinsk Machine Building Plant

COUNTRY OF ORIGIN: U.S.S.R.

OVERALL DIMENSIONS: 5 ft 10 1/2 in. W x 54 ft 11 in. H (179.1 x 1673.9 cm)

WEIGHT: 81,800 lbs (37,100 kg)

KEY ACCOMPLISHMENT(S): Banned by the 1987 Intermediate-Range Nuclear Forces (INF) Treaty, the first international agreement to eliminate an entire class of nuclear weapons

The SS-20, "Pioneer" in Russian, was a mobile, intermediate-range Soviet ballistic missile deployed in the western U.S.S.R. and Siberia beginning in 1976. Capable of striking targets in Western Europe or China, the missiles normally carried three nuclear warheads, each with an explosive force equivalent of 150 kilotons of TNT.

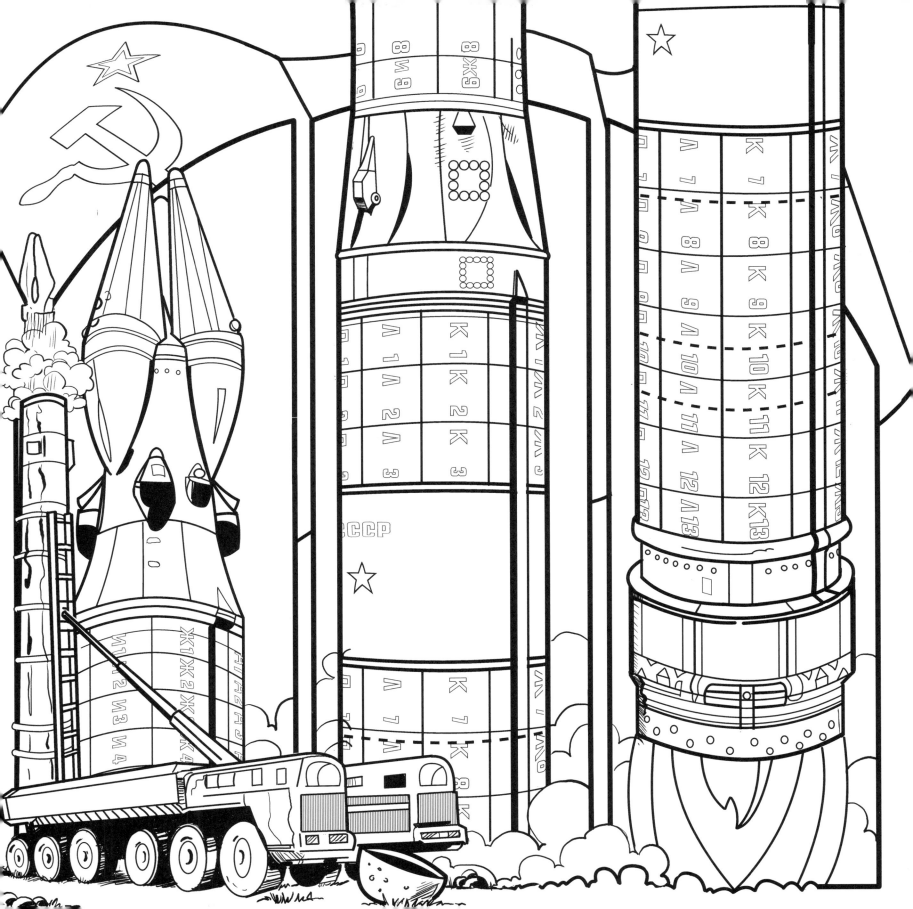

PERSHING II

YEAR: 1981

INVENTOR/MANUFACTURER: Martin Marietta Aerospace

COUNTRY OF ORIGIN: U.S.A.

OVERALL DIMENSIONS: 3 ft 3 5/8 in. dia. x 34 ft 9 5/8 in. H (100.6 x 1060.7 cm)

WEIGHT: 16,451 lbs (7,462 kg)

KEY ACCOMPLISHMENT(S): Banned by the 1987 Intermediate-Range Nuclear Forces (INF) Treaty, the first international agreement to eliminate an entire class of nuclear weapons

The Pershing II was a mobile, intermediate-range U.S. Army ballistic missile deployed in West Germany beginning in 1983. Capable of striking targets in the Western Soviet Union, the missiles carried a single variable-yield nuclear warhead with an explosive force equivalent between 5-80 kilotons of TNT.

SPACE SHUTTLE *DISCOVERY*

YEAR: 1981 (orbital launch)

MANUFACTURER: Rockwell International

COUNTRY OF ORIGIN: U.S.A.

OVERALL DIMENSIONS: 78 ft W x 57 ft H x 122 ft L (24 x 17.7 x 38 m)

WEIGHT: 161,325 lbs (73,176.5 kg)

ROCKET BOOSTER: 418,000 lbf (1,859,356.6 N) per Space Shuttle Main Engine (SSME), of which the orbiter has three, totaling 1,250,000 lbf (5,560,277 N)

KEY ACCOMPLISHMENT(S): The orbiter housed at the Smithsonian, *Discovery*, was the third Space Shuttle to fly into space.

Because *Discovery* flew every kind of mission the Space Shuttle was built to fly, it is a prime symbol of the most active 30 years of U.S. human spaceflight. By the time it was retired from spaceflight in 2011, it was the oldest and most flown orbiter. *Discovery* flew on 39 Earth-orbital missions, spent a total of 365 days in space, and traveled almost 150 million miles (240 million kilometers). It shuttled 184 people into space and back, many more than once. Named for the intrepid sailing ships of the Heroic Age of Exploration, *Discovery* was transferred to the Smithsonian in 2012.

SPACE SHUTTLE MAIN ENGINE

YEAR: 1981 (First Use)

MANUFACTURER: Boeing Rocketdyne

COUNTRY OF ORIGIN: U.S.A.

OVERALL DIMENSIONS: 9 ft 9 in. L x 13 ft 6 in. W x 7 ft 8 in. H (297.2 x 411.5 x 233.7 cm)

WEIGHT: 14,125 lbs (6,407.1 kg)

THRUST: 418,000 lbf (1,859,000 N), liquid oxygen and liquid hydrogen

KEY ACCOMPLISHMENT(S): Although not the most powerful rocket engine ever flown, the Space Shuttle Main Engine (SSME) burned fuel more efficiently than any of its predecessors.

Unlike all of NASA's other liquid-fuel cryogenic rocket engines, the SSME, or Aerojet Rocketdyne RS-25, was reusable. The flights it powered include the first four Shuttle missions, the second Hubble Space Telescope repair mission, the missions that launched the Magellan and Galileo space probes and the John Glenn flight. NASA is planning to continue using the SSME on the Space Shuttle's successor, the Space Launch System (SLS).

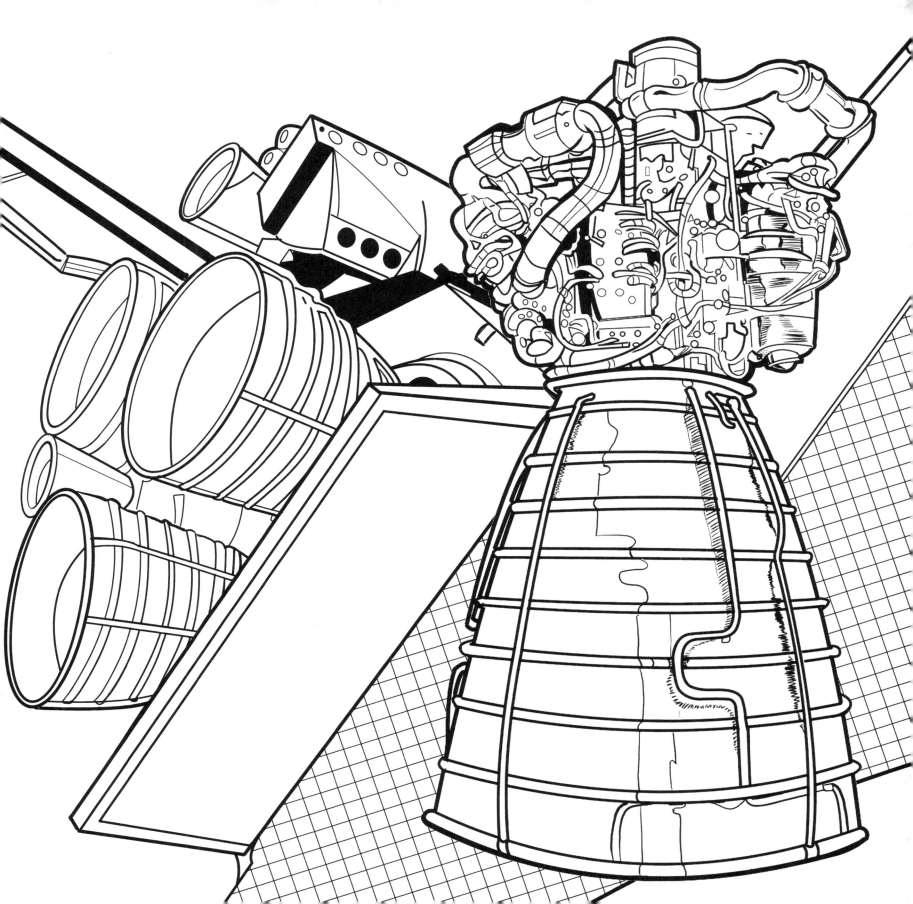

SALLY RIDE'S IN-FLIGHT SUIT, STS-7

YEAR: 1983

INVENTOR/MANUFACTURER: ILC Space Systems

COUNTRY OF ORIGIN: U.S.A.

JACKET DIMENSIONS: 26 1/2 L x 22 W x 3 1/2 in. H (67.3 x 55.9 x 8.9 cm)

KEY TRAIT(S): Size XS, light blue fabric with stitched-on NASA logo and mission patch

KEY ACCOMPLISHMENT(S): American astronaut and physicist Dr. Sally K. Ride wore these clothes during the six-day STS-7 Space Shuttle mission aboard *Challenger* in June 1983, when she became the first U.S. woman and youngest astronaut ever in space.

As a mission specialist on the first five-member Shuttle crew, Ride operated a variety of orbiter systems and experiment payloads. She participated in the launch of two commercial communications satellites and operated the remote manipulator system to maneuver, release and retrieve a free-flying satellite. After flying twice on *Challenger*, she left NASA in 1987 to serve in educational, research and administrative roles.

GUY BLUFORD'S IN-FLIGHT SUIT, STS-8

YEAR: 1983

INVENTOR/MANUFACTURER: ILC Space Systems

COUNTRY OF ORIGIN: U.S.A.

JACKET DIMENSIONS: 26 1/2 L x 41 W x 1 1/2 in. H (67.3 x 104 x 3.8 cm)

KEY TRAIT(S): Size M/L, light blue fabric with attached name tag, NASA logos, and mission patch

KEY ACCOMPLISHMENT(S): Astronaut Guion S. "Guy" Bluford, Jr. wore this suit during the six-day Space Shuttle STS-8 mission aboard *Challenger* in August-September 1983, when he became the first African American in space.

Aerospace engineer, Air Force colonel, and fighter pilot Guion Bluford was selected to become a NASA astronaut in January 1978. As a mission specialist on the STS-8 five-member crew, Dr. Bluford participated in delivering a communications satellite into orbit, operating the Remote Manipulator System and performing experiments. Bluford subsequently served on the crews of the Spacelab D-1 mission in 1985, the largest crew to fly in space, and two missions aboard *Discovery*, STS-39 in 1991 and STS-53 in 1992. With the completion of his fourth flight, Bluford logged over 688 hours in space

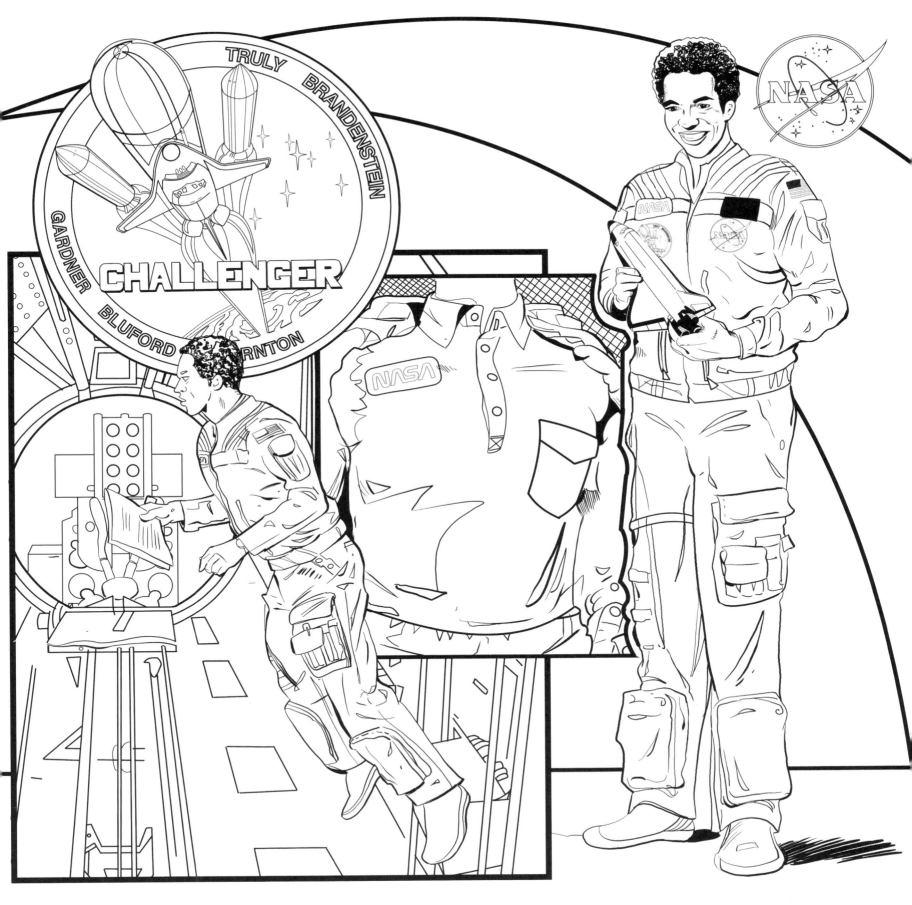

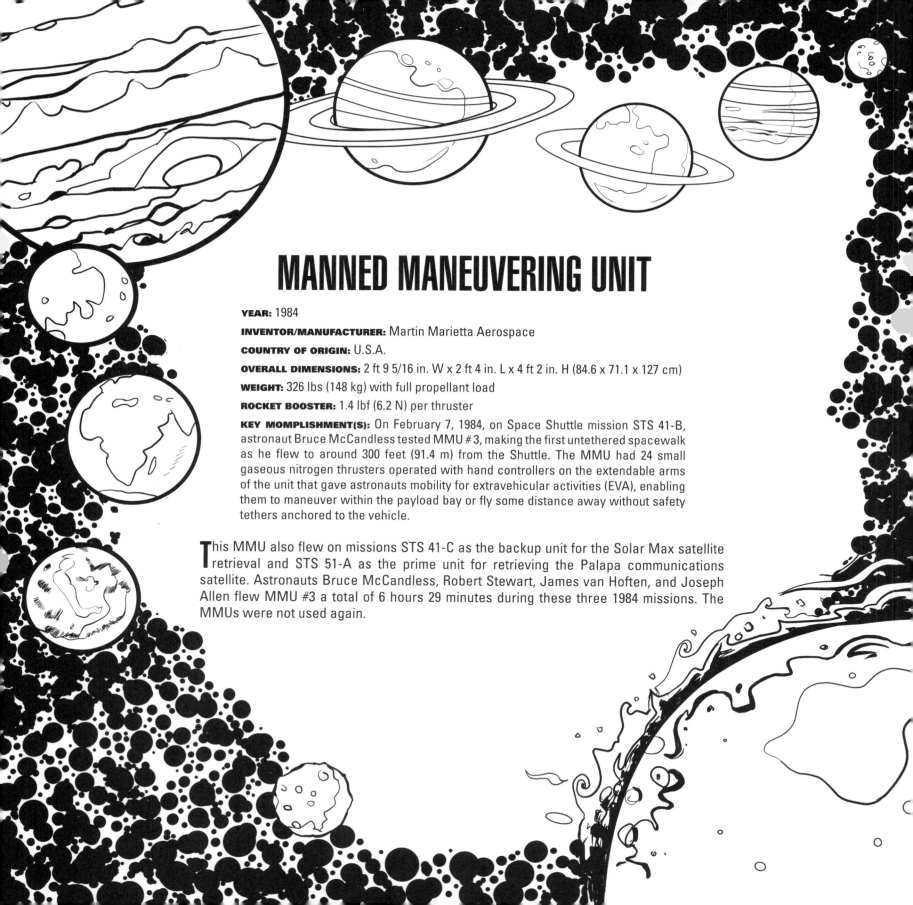

MANNED MANEUVERING UNIT

YEAR: 1984

INVENTOR/MANUFACTURER: Martin Marietta Aerospace

COUNTRY OF ORIGIN: U.S.A.

OVERALL DIMENSIONS: 2 ft 9 5/16 in. W x 2 ft 4 in. L x 4 ft 2 in. H (84.6 x 71.1 x 127 cm)

WEIGHT: 326 lbs (148 kg) with full propellant load

ROCKET BOOSTER: 1.4 lbf (6.2 N) per thruster

KEY MOMPLISHMENT(S): On February 7, 1984, on Space Shuttle mission STS 41-B, astronaut Bruce McCandless tested MMU #3, making the first untethered spacewalk as he flew to around 300 feet (91.4 m) from the Shuttle. The MMU had 24 small gaseous nitrogen thrusters operated with hand controllers on the extendable arms of the unit that gave astronauts mobility for extravehicular activities (EVA), enabling them to maneuver within the payload bay or fly some distance away without safety tethers anchored to the vehicle.

This MMU also flew on missions STS 41-C as the backup unit for the Solar Max satellite retrieval and STS 51-A as the prime unit for retrieving the Palapa communications satellite. Astronauts Bruce McCandless, Robert Stewart, James van Hoften, and Joseph Allen flew MMU #3 a total of 6 hours 29 minutes during these three 1984 missions. The MMUs were not used again.

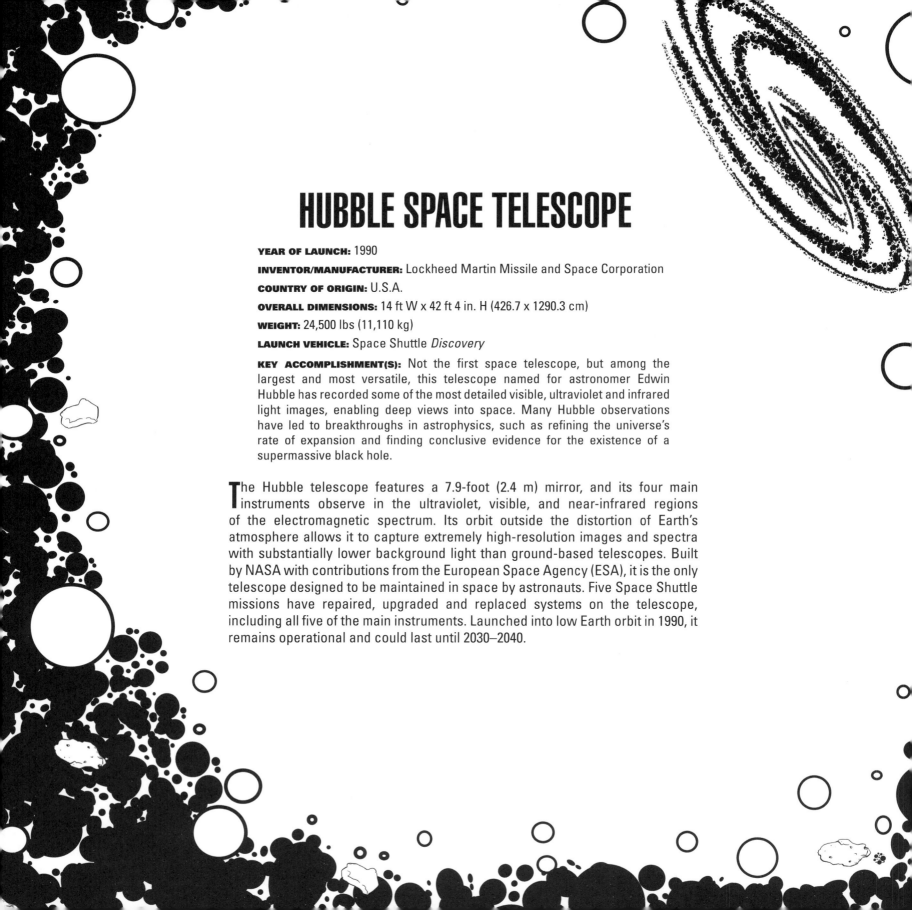

HUBBLE SPACE TELESCOPE

YEAR OF LAUNCH: 1990

INVENTOR/MANUFACTURER: Lockheed Martin Missile and Space Corporation

COUNTRY OF ORIGIN: U.S.A.

OVERALL DIMENSIONS: 14 ft W x 42 ft 4 in. H (426.7 x 1290.3 cm)

WEIGHT: 24,500 lbs (11,110 kg)

LAUNCH VEHICLE: Space Shuttle *Discovery*

KEY ACCOMPLISHMENT(S): Not the first space telescope, but among the largest and most versatile, this telescope named for astronomer Edwin Hubble has recorded some of the most detailed visible, ultraviolet and infrared light images, enabling deep views into space. Many Hubble observations have led to breakthroughs in astrophysics, such as refining the universe's rate of expansion and finding conclusive evidence for the existence of a supermassive black hole.

The Hubble telescope features a 7.9-foot (2.4 m) mirror, and its four main instruments observe in the ultraviolet, visible, and near-infrared regions of the electromagnetic spectrum. Its orbit outside the distortion of Earth's atmosphere allows it to capture extremely high-resolution images and spectra with substantially lower background light than ground-based telescopes. Built by NASA with contributions from the European Space Agency (ESA), it is the only telescope designed to be maintained in space by astronauts. Five Space Shuttle missions have repaired, upgraded and replaced systems on the telescope, including all five of the main instruments. Launched into low Earth orbit in 1990, it remains operational and could last until 2030–2040.

ROVER, MARIE CURIE, MARS PATHFINDER ENGINEERING TEST VEHICLE

YEAR: 1997

COUNTRY OF ORIGIN: U.S.A.

INVENTOR/MANUFACTURER: Jet Propulsion Laboratory (JPL), California Institute of Technology

OVERALL DIMENSIONS: 2 ft 3 1/2 in. L × 1 ft 5 1/2 in. W × 1 ft 1/2 in. H (69.8 × 44.4 × 31.7 cm)

WEIGHT: 1,982 lbs (899 kg)

LAUNCH VEHICLE: Delta II

KEY ACCOMPLISHMENT(S): The Marie Curie rover was the flight spare for the Sojourner rover. During Sojourner's activities on Mars, engineers operated Marie Curie in the same movements in a Mars-like test area at the Jet Propulsion Laboratory (JPL) in California.

Mars Pathfinder was the first spacecraft to land on the surface of the red planet since the Viking mission in 1976. On reaching Mars on July 4, 1997, the spacecraft entered the planet's thin atmosphere. Once on the surface, the protective aeroshell unfolded to provide three flat platforms and ramps, one of which held a rover (Sojourner). The Sojourner rover traveled down one of the ramps and proceeded to take close up images of the surface using two color cameras on the front and a black and white camera on the rear. The rover also featured a rear-mounted Alpha Proto X-ray Spectrometer that provided bulk elemental composition data on surface soils and rocks.

STARDUST RETURN CAPSULE

YEAR: 1999

COUNTRY OF ORIGIN: U.S.A.

WEIGHT: 101 lbs (45.8 kg)

LAUNCH VEHICLE: Delta II

INVENTOR/MANUFACTURER: Lockheed Martin Missile and Space Corporation

OVERALL DIMENSIONS: 2 ft 3 1/2 in. L × 1 ft 5 1/2 in. W × 1 ft 1/2 in. H (86.4 x 157.5 x 81.3 cm)

KEY ACCOMPLISHMENT(S): Stardust was the first U.S. space mission dedicated solely to returning extraterrestrial material from beyond the Moon.

Stardust launched in February 1999 on a 3 billion-mile roundtrip to rendezvous with Comet Wild 2, capture comet and interstellar dust and return a capsule bearing these primordial solar system "treasures" for analysis here on Earth. Seven years later, the journey ended with the capsule streaking across the sky to a successful landing on U.S. soil in January 2006. Since then, the dust samples have gone to laboratories around the world for scientists to study the chemical composition of the comet and its signature of the early solar system. Stardust accomplished the first U.S. robotic sample return mission beyond the Moon and the first collection of comet material for study on Earth. Its return marked the fastest atmospheric entry of a human-made object at about 29,000 miles per hour.

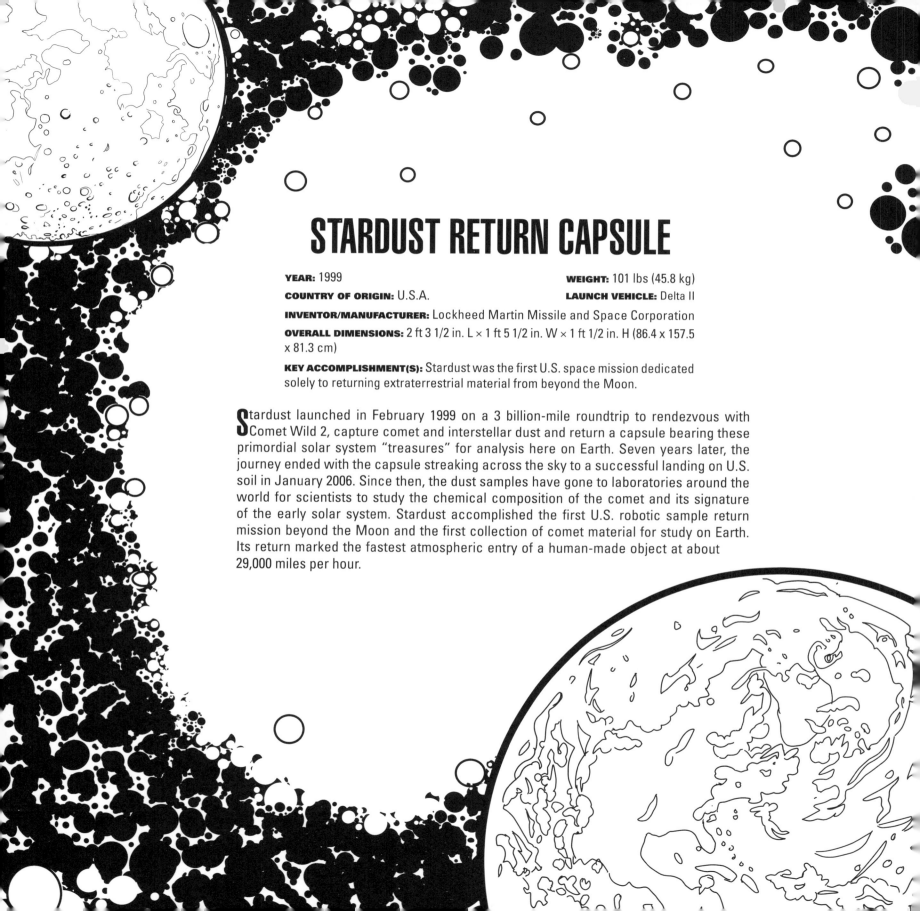

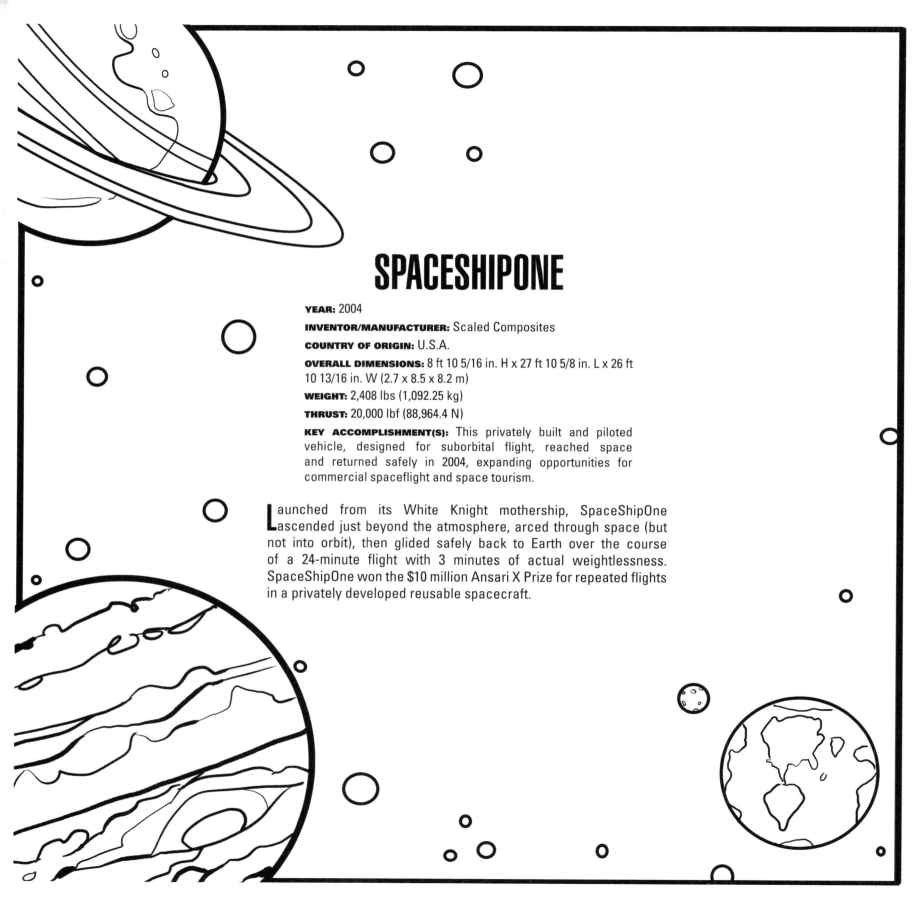

SPACESHIPONE

YEAR: 2004

INVENTOR/MANUFACTURER: Scaled Composites

COUNTRY OF ORIGIN: U.S.A.

OVERALL DIMENSIONS: 8 ft 10 5/16 in. H x 27 ft 10 5/8 in. L x 26 ft 10 13/16 in. W (2.7 x 8.5 x 8.2 m)

WEIGHT: 2,408 lbs (1,092.25 kg)

THRUST: 20,000 lbf (88,964.4 N)

KEY ACCOMPLISHMENT(S): This privately built and piloted vehicle, designed for suborbital flight, reached space and returned safely in 2004, expanding opportunities for commercial spaceflight and space tourism.

Launched from its White Knight mothership, SpaceShipOne ascended just beyond the atmosphere, arced through space (but not into orbit), then glided safely back to Earth over the course of a 24-minute flight with 3 minutes of actual weightlessness. SpaceShipOne won the $10 million Ansari X Prize for repeated flights in a privately developed reusable spacecraft.

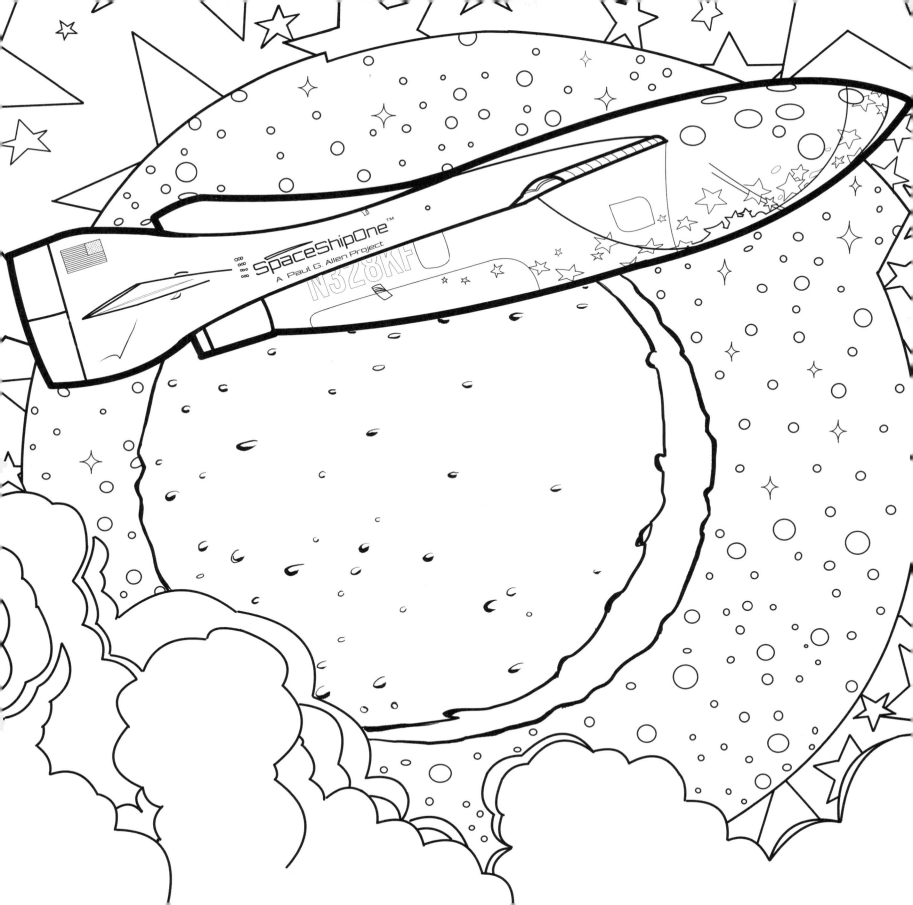

NEW HORIZONS

YEAR: 2006

INVENTOR/MANUFACTURER: John Hopkins University Applied Physics Laboratory (APL), Southwest Research Institute (SwRI)

COUNTRY OF ORIGIN: U.S.A.

OVERALL DIMENSIONS: 7.2 ft L x 6.9 ft W × 8.9 ft H (2.2 × 2.1 × 2.7 m)

WEIGHT: 1,054 lbs (478 kg)

THRUST: 200 lbf (890 N)

LAUNCH VEHICLE: Atlas V 551

KEY ACCOMPLISHMENT(S): New Horizons is the first spacecraft to explore Pluto, its moons, and the icy bodies of the Kuiper Belt in the outer solar system. The fifth space probe to achieve the escape velocity needed to leave the solar system, it was also the fastest man-made object ever launched from Earth at its time of launch.

Just over a year after its January 2006 launch, New Horizons conducted a Jupiter flyby, thus gaining further acceleration to reach Pluto in July 2015. In August 2018, the spacecraft's ultraviolet imaging spectrometer, Alice, confirmed the existence of a "hydrogen wall" at the outer edges of the solar system. New Horizons' payload included scientific instruments to map the surface geology and composition of Pluto and its three moons, investigate Pluto's atmosphere, measure the solar wind and assess interplanetary dust and other particles. It also carried several souvenirs from Earth, including some of the remains of Clyde Tombaugh (1906-1997), discoverer of Pluto, and a piece of SpaceShipOne.

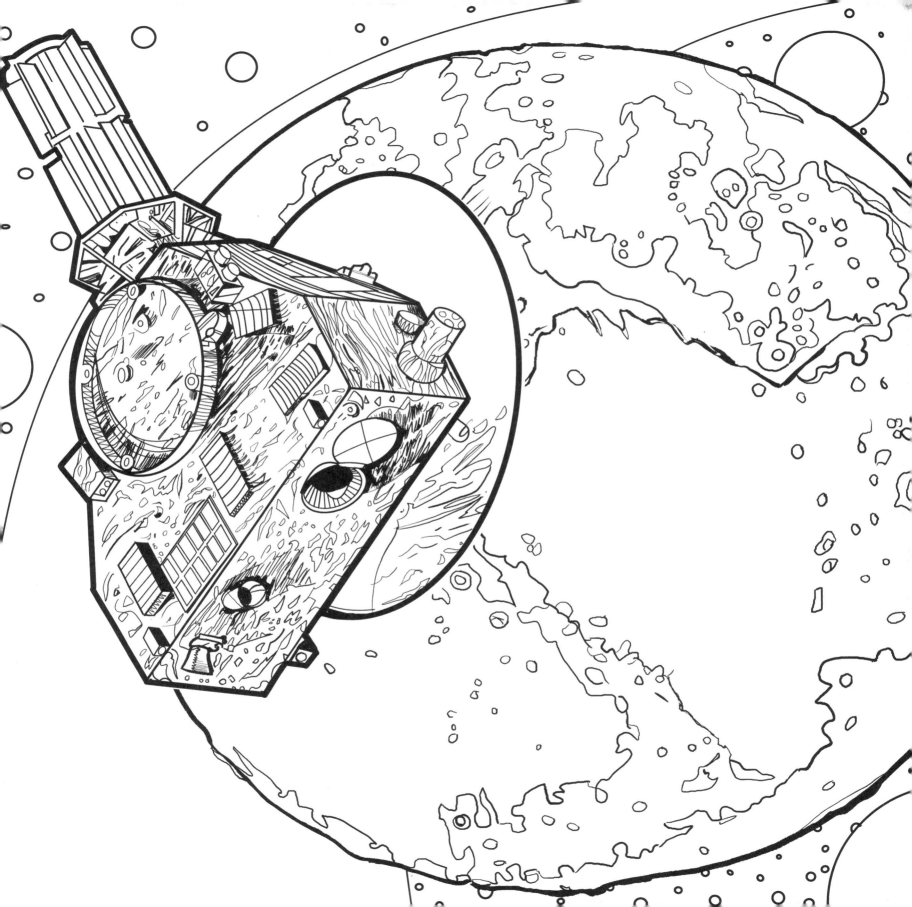

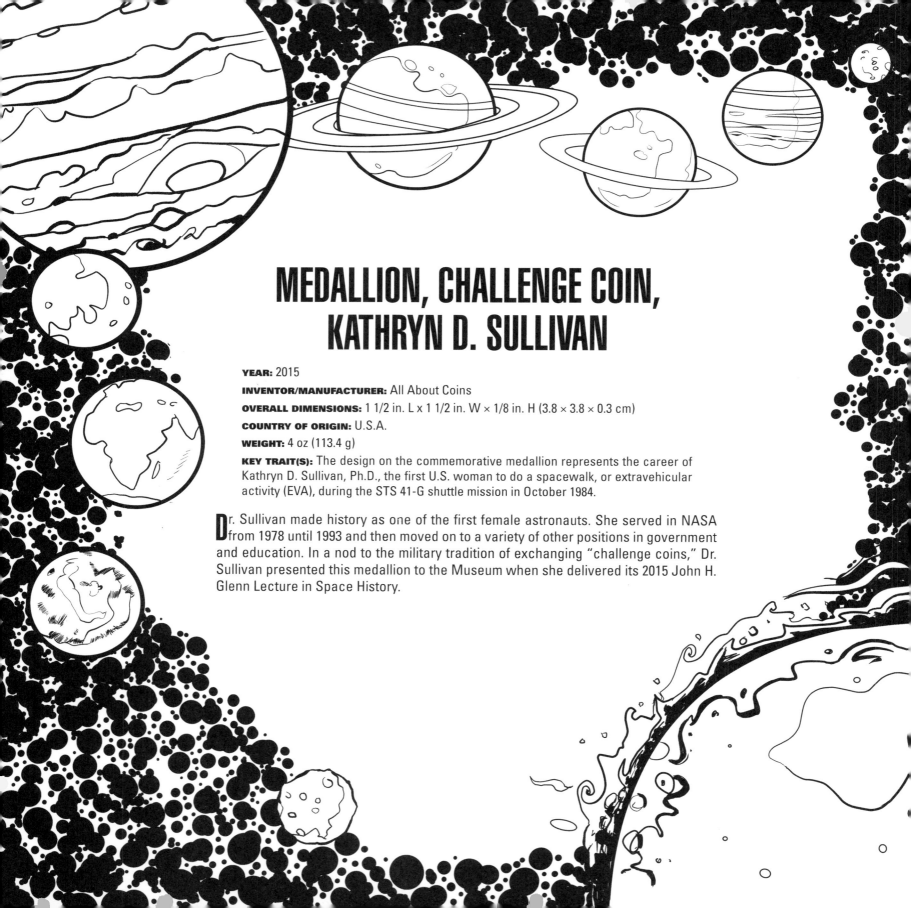

MEDALLION, CHALLENGE COIN, KATHRYN D. SULLIVAN

YEAR: 2015

INVENTOR/MANUFACTURER: All About Coins

OVERALL DIMENSIONS: 1 1/2 in. L x 1 1/2 in. W × 1/8 in. H (3.8 × 3.8 × 0.3 cm)

COUNTRY OF ORIGIN: U.S.A.

WEIGHT: 4 oz (113.4 g)

KEY TRAIT(S): The design on the commemorative medallion represents the career of Kathryn D. Sullivan, Ph.D., the first U.S. woman to do a spacewalk, or extravehicular activity (EVA), during the STS 41-G shuttle mission in October 1984.

Dr. Sullivan made history as one of the first female astronauts. She served in NASA from 1978 until 1993 and then moved on to a variety of other positions in government and education. In a nod to the military tradition of exchanging "challenge coins," Dr. Sullivan presented this medallion to the Museum when she delivered its 2015 John H. Glenn Lecture in Space History.

Smithsonian

SPACECRAFT

Coloring Book

ILLUSTRATED BY
John Pirtel

IDW®